T0234353

Synthesis Lectures on Engineering, Science, and Technology

The focus of this series is general topics, and applications about, and for, engineers and scientists on a wide array of applications, methods and advances. Most titles cover subjects such as professional development, education, and study skills, as well as basic introductory undergraduate material and other topics appropriate for a broader and less technical audience.

Kersten Heins

Trusted Cellular IoT Devices

Design Ingredients and Concepts

Kersten Heins
Munich, Germany

ISSN 2690-0300   ISSN 2690-0327 (electronic)
Synthesis Lectures on Engineering, Science, and Technology
ISBN 978-3-031-19665-2   ISBN 978-3-031-19663-8 (eBook)
https://doi.org/10.1007/978-3-031-19663-8

© The Editor(s) (if applicable) and The Author(s), under exclusive license to Springer Nature Switzerland AG
2022
This work is subject to copyright. All rights are solely and exclusively licensed by the Publisher, whether the whole
or part of the material is concerned, specifically the rights of translation, reprinting, reuse of illustrations, recitation,
broadcasting, reproduction on microfilms or in any other physical way, and transmission or information storage
and retrieval, electronic adaptation, computer software, or by similar or dissimilar methodology now known or
hereafter developed.
The use of general descriptive names, registered names, trademarks, service marks, etc. in this publication does
not imply, even in the absence of a specific statement, that such names are exempt from the relevant protective
laws and regulations and therefore free for general use.
The publisher, the authors, and the editors are safe to assume that the advice and information in this book are
believed to be true and accurate at the date of publication. Neither the publisher nor the authors or the editors give
a warranty, expressed or implied, with respect to the material contained herein or for any errors or omissions that
may have been made. The publisher remains neutral with regard to jurisdictional claims in published maps and
institutional affiliations.

This Springer imprint is published by the registered company Springer Nature Switzerland AG
The registered company address is: Gewerbestrasse 11, 6330 Cham, Switzerland

Preface

On this planet, everybody understands the **value of trust**. In fact, trust is a major prerequisite for every successful cooperation and provides mutual confidence that expectations will be met by the other party. Trust might be based on prior experience or evidence, or it is just a soft assessment. This is fundamental for human business partners, but what about machines? What makes you certain that a connected electronic device will work as expected? How can we trust machines?

As a matter of fact, the new Internet of Things (IoT) approach will be able to substitute many human interactions by automated processes - offering increased efficiency, better service quality or user convenience. Potential IoT benefits have been fully understood all over the world and related value propositions will continue to fuel the IoT success story, but only if IoT solutions offer an unquestioned level of **reliability and confidence**. Without a decent level trust in IoT, users will turn away and will prefer to return to the traditional way of working.

IoT service providers have started to recognize that each failure and each security breach might irrevocably torpedo their business. Reports about malfunction or confidentiality leaks of IoT applications will destroy user confidence and cause severe commercial damage and/or legal consequences. Governments are aware of this topic and already have started to put IoT-related laws in place. Consequently, each business owner and the whole IoT industry will have to ensure that their solutions are trustworthy and well accepted by users.

For this purpose, implementation of appropriate security measures is crucial to minimize risks and to ensure reliable and trustworthy operation of IoT ecosystems. "IoT Security" is the magic spell. In fact, **IoT security is the key enabler for a trusted Internet of Things**, because it helps to build a trusted business relationship between service providers and customers. Security evaluations of IoT devices and certificates by independent third parties can support this confidence-building process.

Munich, Germany   Kersten Heins

About This book

According to their nature, IoT sensors observing remote locations are not physically controlled by operators, and typically they work at unattended or unprotected locations. As such, remote **IoT devices are exposed and vulnerable** against attacks or misuse. In most cases, IoT device are the weakest point of an IoT ecosystem and the most attractive target for IoT attackers to reach their goals most efficiently. This risk requires an extra level of attention during the development of an IoT device. Consequently, a design engineer should have a good understanding of risks, potential attacks and entry doors for security breaches and data leaks of an IoT device.

This book explains how to identify application-specific vulnerabilities and how to address them most efficiently by design of an IoT device. This book explains all major ingredients for trusted IoT devices and provides direction for proper selection. As a reference technology for secure IoT connectivity, cellular networks (LTE-M and NB-IoT) have been used, but most of the book contents is also valid for IoT projects using other wireless networks like WiFi or LoRaWAN or SigFox. In particular, all explanations of IoT risk assessment, security requirements, protection countermeasures as well as introductions of design ingredients and concepts are applicable to different network technologies.

Chapter 1 is called "Introduction and Scope". It explains the architecture of an IoT ecosystem, and which IoT security aspects are relevant for trustworthy operation, overall user acceptance, and commercial success of an IoT application. It outlines the benefits of cellular networks for IoT connectivity and explains why IoT security considerations should focus on device design.

Chapter 2 is called "Challenges and Objectives". It explains how various threats can jeopardize the overall trustability of IoT ecosystems. IoT project owners will learn, which risks might lead to serious damage and how to assess related threads and attack scenarios. Based on application-specific requirements, and in combination with legal obligations, a balanced level of countermeasures will have to protect IoT deployments and related business models. A smart meter reference project has been used for illustration of a risk assessment process. In addition, this chapter explains why an independent IoT security evaluation and product certification process is important.

Chapter 3 is called "Cryptographic Toolkit". It starts with some security fundamentals and explains major cryptographic methods for protection of IoT deployments such as digital signatures, MACs and PKIs. Then, a selection of popular useful protocols like TLS or the Diffie-Hellman key exchange are covered.

Chapter 4 is called "Ingredients for Secure Design". It explains how designers can benefit from standard security products and outlines standard countermeasures against physical and logical intrusion as well as typical solutions for secure firmware upgrade, integrity checking or secure boot. Then, it provides a market snapshot of commercial off-the-shelf solutions for IoT device design and how to handle production lifecycles.

Chapter 5 is called "Device Design". It provides a list of universal design tips, a concept for active intrusion sensing and for an ultra-secure IoT device with built-in integrity protection and secure boot - based on a secure element.

Finally, a comprehensive **Glossary** provides explanations of common terms and acronyms related to technical IoT, cellular network and security topics.

Contents

About the Author

Kersten Heins is a passionate technical consultant and content creator for IoT ecosystems, focusing on device design and embedded security (see www.iot-chips.com). After his graduation he spent 30 years in the industrial computing and semiconductor industry as a design engineer and in various marketing positions.

List of Figures

Introduction and Scope

1

There is no doubt that the "Internet of Things" (IoT) will change many aspects of our daily lives. Commercial expectations are high, and world-wide IoT industry and service providers are enthusiastic about forecasted market numbers and corresponding demand for billions of IoT devices. IoT success is fueled by a pervasive 24/7 Internet coverage while data transmission cost is continuously decreasing. Based on advanced machine-to-machine (M2M) networking solutions, IoT has been established as a new approach how to address **existing** use cases and improve existing processes—e.g., by replacing human interaction by sensors and actuators. But the **IoT approach** also creates **new** appealing applications for consumers as well as for industrial users.

IoT is following the concept of "edge computing", i.e., it manages end devices at remote locations to monitor and preprocess local data and perform local actions. While the remote IoT device performs more or less autonomously most of the time, it is connected to the Internet and supervised by a central IoT application through the Internet. This way, simple "things" become **smart IoT devices**.

As a matter of fact, IoT is an umbrella term for an endless list of target applications. Most of them are **monitoring and analyzing conditions**, aiming at an improved process **efficiency** resp. **reduction of operational cost** esp. for industrial or business-to-business (B2) use cases. Other IoT applications are helping to make everyday tasks **more convenient and to improve quality of life**. Many new IoT services and applications will affect our daily lives—either directly or indirectly.

Examples:

- A connected vending machine can be always operational, and always filled up with goods.

© The Author(s), under exclusive license to Springer Nature Switzerland AG 2022
K. Heins, *Trusted Cellular IoT Devices*, Synthesis Lectures on Engineering, Science, and Technology, https://doi.org/10.1007/978-3-031-19663-8_1

- Predictive maintenance techniques enable machines to determine service needs by themselves, decrease machine downtimes, and eliminate fixed maintenance schedules at the same time.
- Smart homes and offices can save energy costs by controlling the electricity or temperature automatically or remotely, and they can offer better security by constant surveillance and take proactive action in case of a housebreaking attempt.
- Smart automobiles can request assistance if required or assist in monitoring vehicle speed based on traffic.
- eHealth services can improve health care by monitoring patients and remotely administering medication to them.

As a consequence, deployment of IoT applications concerns all of us and increasingly attracts public interest and also governmental attention.

1.1   Social Impact and Public Relevance

IoT applications have already started to penetrate our homes, production facilities, public areas, transportation vehicles, warehouses, etc. and obviously has a significant **social impact**. For example, IoT technology causes a positive social perception whenever remote monitoring is used for safety purposes or reduction of air pollution. But on the other hand, social feedback might turn into negative if used for remote-controlled observation equipment which is potentially violating personal rights. Social impact depends on IoT application and user roles. In general, there is no doubt that IoT technology will generate a lot of positive feedback from individuals who take advantage of IoT benefits (e.g., cost reduction).

IoT technology enables management and automation of remote processes, but at the same time it eliminates need for human local interaction—and related jobs. Cost reduction effects are beneficial for business owners, but might frustrate other members of our society, esp. employees. In an IoT-enabled world, different professional skills will be required. In general, IoT tends to create qualified operator jobs, but at the same time also replaces lower profile labor and travel by automation and remote control IoT features. In these cases, resulting social impact depends on which position you are looking at it.

A universal social concern is **privacy**, i.e., eavesdropping of personal or private data. By nature, IoT applications are transmitting remote data from/to a specific location which is associated to an individual or responsible user. Sensitivity of transmitted data depends on the application, e.g., it could be real-time information about

- electricity consumption (smart meter)
- audio/video access (set-top box)
- heart rate (eHealth)

- lighting switch on/off commands (smart home/office)
- surveillance video (security).

In any case, transmission of this kind of personal data must be protected against unauthorized access in order to ensure privacy. In fact, intercepted contents itself might be sensitive (e.g., personal health data), but even less confidential data might be used to create **user profiles** containing individual presence timestamps or consumption data allows to determine user habits, shopping preferences, etc. Protection against collection and processing of personal data is a fundamental citizen right specified by national laws of many countries worldwide (**data privacy regulations**). Affected individuals will have to explicitly agree, otherwise it will be not allowed at all. In fact, we find several dedicated national **IoT cybersecurity laws** which are asking manufacturers to implement reasonable security features to prevent unauthorized access. We will take a close look at this subject later in a dedicated Sect. 2.9.

In general, protection of privacy is part of the overall **data security requirements** of an IoT ecosystem such as:

- manage access control and
- ensure data confidentiality, integrity, and authenticity.

In an effort to meet IoT data security requirements, manufacturers and service providers will have to put a reasonable set of technical and organizational countermeasures in place. Suitable cryptographic techniques and associated technical toolboxes will be described later in Chaps. 3 and 4.

In any case, **reliable and trustworthy IoT applications** will boost user confidence to let IoT processes handle big data and critical processes in the background. If not, people will tend to object. Very obviously, a well-considered and secure IoT system concept and proper implementation of each of its elements are key aspects for a successful rollout of an IoT application.

1.2 IoT Ecosystem

"IoT" stands for a concept rather than a single application. Instead, the term IoT is used as an umbrella term for many different applications and associated technologies. But from a top-level point of view, they share a set of common characteristics and enablers which are illustrated in Fig. 1.1.

In general, an IoT application is monitoring **multiple remote locations** (usually a large number) by sensing local parameters or just detecting local events, e.g., presence of an object. For this purpose, an IoT application is working with **IoT devices** sending local IoT which are reflecting relevant aspects of an observed environment or object.

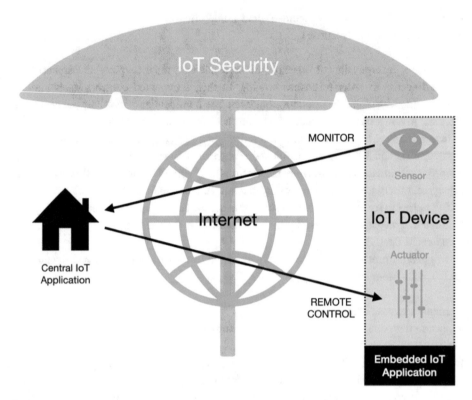

Fig. 1.1 IoT system overview (for a single IoT device)

Figure 1.1 is outlining an IoT system with just one sample IoT device. Raw IoT data set is being transmitted to a **central IoT application server** which is consolidating all IoT data streams received from connected IoT devices. On server side (or a cloud service), further data analysis and application-specific data processing will be performed and might result in actions to be performed locally ("remote control").

In short, an **IoT ecosystem** is formed from a set of elements which can be categorized like this:

1. Sensors and actuators and embedded application (device)
2. Network connectivity (communication channel)
3. Central application and data analytics (cloud software and services)
4. Security.

Security measures provide protection against attacks and unintended misuse of an IoT application. Very often, an IoT application is competing against use cases which are attended or supervised by human operators. As such, it might qualify as an efficient alternative, if appropriate **IoT security** means have been implemented in an effort to ensure a trustworthy and reliable IoT application with remote frontend devices which are typically unattended. In general, IoT security affects all elements of the ecosystem, i.e., the device, the communication channel and cloud-based software. As a system concept, IoT Security has to be tailored to each use case.

Unfortunately, there is no standard ecosystem or platform for all kind of IoT use cases. Instead, each IoT project requires a business- and application-specific selection of IoT building blocks and a concept how to connect them. Elements of an IoT ecosystem are offered by a large number of suppliers (see Fig. 1.2), each of them providing products or services in their area(s) of expertise. IoT project managers and business owners will have to short-list suitable candidates for all relevant building blocks of a suitable IoT ecosystem. In order to offer an attractive one-stop-shopping experience to their customer, many hardware suppliers are bundling their core products (e.g., a network interface module) with corresponding cloud-based services, e.g., an IoT device management tool or subscription plans for network connectivity. Many of these companies are mentioned in the course of this book, esp. in Chap. 4. Another starting point are the following websites which are introducing a lot of IoT service providers and suppliers:

- https://www.iot-directory.com
- https://www.iotglobalnetwork.com/list/companies
- https://www.iotavenue.com/iot-company-directory.

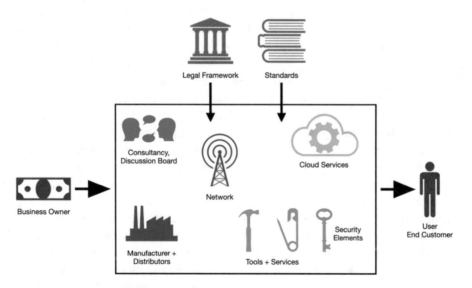

Fig. 1.2 IoT ecosystem—stakeholders

On top of collaboration with suitable business partners, IoT projects will have to meet governmental obligations which are often aiming at personal data protection or fraud prevention (see Sect. 2.9). In addition, some IoT products will have to proof compliance with dedicated security regulations for specific target markets, esp. for national projects of public interest, e.g., electricity meters (see Sect. 2.10).

1.3   What is an "IoT Device" ?

All local activities are handled by smart IoT devices. An IoT device is a dedicated embedded system which is acting on behalf of the IoT application and located close to data sources of interest, i.e., at the far end of an IoT application ("edge computing"). An IoT device is a remote-controlled MCU-based system which is running independently and does not require any local human interaction and often does not even offer any local user interface. Instead, the IoT device is a managed by **embedded IoT software** which is working in cooperation with the central IoT application, i.e., the overall IoT application is separated into a local part running on the IoT device and a server part.

Figure 1.3 is illustrating the main functional elements of an IoT device. The embedded IoT software is mainly responsible for hardware control and typically does not require much computing power, i.e., a simple 1-dollar microcontroller can handle it. This part of the IoT application is fixed and has been stored permanently in a non-volatile memory during production, therefor also called **IoT firmware**. The IoT firmware starts after right after a IoT device has been powered up or after a device reset and works offline, but supervised and interruptible by the IoT server, if needed. Optionally and if required, it can also be updated by the IoT server via network (FOTA = firmware-over-the-air). The device firmware determines the hardware flow, i.e., order and type of local actions, and manages other device components via dedicated interfaces (e.g., GPIO pins or serial I^2C or UART interfaces accordingly. For this purpose, each local peripheral interaction needs to be converted into dedicated I/O control and data commands to be submitted according to the interface protocol specifications. The IoT firmware has to communicate with the **network interface** module and interact with peripheral IoT **sensors and actuators**. Besides these components for remote sensing and control, Internet connectivity another fundamental IoT requirement. The network interface is a central element and one of the most complex building blocks of the IoT device design. Usually, it is implemented as a dedicated network subsystem (network interface module) with an integrated MCU, memory, RF-section incl. antenna interface and SIM card interface. Network interface modules are quite expensive (around 10 USD or more), but they handle lower network layers independently and offer a high-level API which allows the IoT firmware to delegate adjustment of network parameters, establish connections and handle communication sessions. In addition, Internet sockets and protocols as well as security features are providing significant added value to the core function of the network interface.

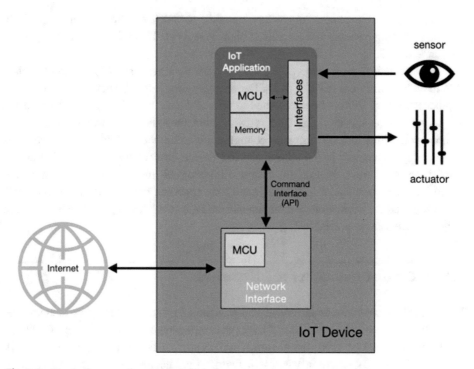

Fig. 1.3 Block diagram of a generic IoT device

The local IoT application can also perform simple calculations and take basic decisions, e.g., if a certain pre-defined threshold has been exceeded and requires submitting an alert to the central IoT application (example: local ambient temperature is exceeding a certain, predefined range).

A fundamental truth is that different IoT applications are following different objectives and will be looking for different local conditions and require different edge computing capabilities. In fact, each IoT application requires tailored sensors and actuators, a suitable microcontroller plus memory, matching connectivity parameters (range, latency, data rate, etc.) and additional functional elements or services. As a consequence, there is **no off-the-shelf general-purpose IoT device** available on the market. For IoT application owners it would be beneficial to use a configurable standard IoT platform offering all required features and configure according to custom requirements. This approach would be able to reduce own development activities and increase time-to-market. But this approach would be too expensive because, each customer would have to pay for redundant components (e.g., a temperature sensor) or functions (e.g., battery management) which are not required for every IoT application. Consequently, at least for large-scale IoT installations, unit price limitations will justify development of a tailored **custom IoT device with optimized characteristics**. Fortunately, designers can benefit from standard lower-granularity

IoT functional elements such as network interface modules, sensors, antennas, etc. incl. software building package and associated high-level APIs. This subject will be explained later in Sect. 4.3.

From a technical perspective, it is possible to set up a **generic IoT platform** which is able to address different IoT use cases. For example, single-board computer systems like popular Raspberry Pi have been designed to offer ultimate flexibility and can be adjusted for an application-specific hard- and software implementation. Hardware extensions in combination with a suitable software configuration and embedded application programs are allowing a customer to put a complete IoT device in place with lowest effort and in a short period of time. Although a custom Raspberry Pi implementation would be too expensive for a commercial IoT deployment, this approach makes perfect sense for IoT feasibility studies, evaluation of components or as a mock-up, e.g., for demonstration purposes (see sample project in [1]).

1.4   Cellular and Other IoT Networks

Usually, IoT connectivity requirements can be handled by standard network technologies available on the market. For some IoT target environments (e.g., for production facilities), fixed networks can be used, but for most IoT deployments use of a wireless network is more efficient. By nature, a wireless network offers flexible connectivity to arbitrary device locations which are within reach of a network access point. For mobile applications resp. use cases with moving resp. movable IoT devices, fixed networks cannot be used anyway (refer to [1] for a comprehensive overview of IoT use cases). But even for stationary IoT devices, a wireless option might be useful in addition to an existing network infrastructure. This is particularly beneficial whenever the target deployment area is difficult to control from an IoT service provider point of view, e.g., a private site or an office building. In this case, the IoT device can be managed via wireless connection in a fully independent manner.

Each commercial IoT project is based on its own business conditions, and some technical parameters might be fixed. For example, if you follow an invitation to a public tender for a national smart meter project, used network technology and infrastructure will be predefined, i.e., not negotiable. But in most cases and from a technical point of view, various wireless technologies will be able to meet requirements of a specific IoT application, not just one. Figure 1.4 provides an overview of popular wireless technologies which might be used for IoT projects.

Key criteria for selection of the most appropriate network technology for a specific IoT application are

- **data rate** (transmission speed),
- **range** (distance from network access point to deployed IoT device) and

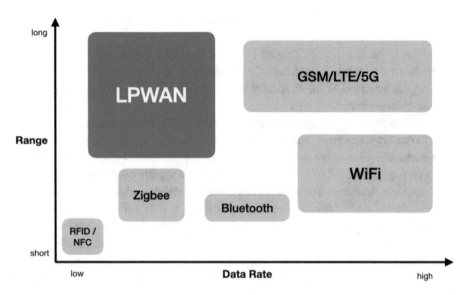

Fig. 1.4 Wireless IoT networks

- **power consumption** (of the IoT device).

NFC (near-field communication) with very short transmission range is used for contactless tokens rather than for IoT devices requiring larger distance data transmission of several meters or even kilometers. **Bluetooth** and **Zigbee** networks offer ranges of up to 100 m and can be used locally as a secondary network in combination with a gateway device offering Internet access. Popular **Wi-Fi** networks are based on the IEEE 802.11 family of standards, and they are commonly used for **local area networking** (LAN) allowing nearby digital devices to access the Internet. A Wi-Fi access point usually has a range of about 20 m indoors and maybe up to 100 m outdoors. Wi-Fi can achieve speeds of several Gbit/s which is a key advantage for applications asking for very high data rates (which is not a typical IoT requirement). Standard cellular networks like **LTE** are providing data rates of as much as several Mbit/s and access points with a coverage area of several kilometers ("cells").

A new category of wireless network technologies is called **LPWAN** (= "Low-Power Wide Area Network") which stands for long-range communication at comparably low data rates. LPWANs are cellular technologies NB-IoT or LTE-M as well as competing technologies **LoRaWAN** and **Sigfox** which are no cellular networks and operate in the so-called **unlicensed frequency spectrum**. For example, the Wi-Fi IEEE 802.11 standards provides several distinct radio frequency ranges for use in Wi-Fi communications (esp. 2.4 GHz and 5 GHz) which are free to use—without a license but according to a specific set of rules. The rules may correspond to a specific access protocol to ensure fairness, restrictions on power levels, etc.

In general, the use of radio frequencies is regulated by national authorities. Cellular network technologies which have been developed under the worldwide 3GPP [2] umbrella are using the so-called **licensed spectrum** (see [3] for further information about radio spectrum management).

This means that network operators have acquired the right to use dedicated parts of the public frequency space from national or regional authorities—under the condition to provide a connectivity service to the public. Typically, for governmental **spectrum allocation** an auction process is used to sell the rights to transmit signals over specific bands of the electromagnetic spectrum. The idea was to put a worldwide cell-based and cross-border infrastructure for mobile communication in place, i.e., mainly for transmission of voice.

Originally, this cellular network approach has been developed as GSM (= "Global System for Mobile Communications") technology for digital mobile phones—to be used across national borders. International cooperation work started 1982 in Europe, and first GSM networks were implemented from 1992 onwards. Over time, technical specification work moved from European ETSI to **3GPP** (3GPP = "Third Generation Partnership Project"). 3GPP [2] is the global standardization organization behind the evolution and maintenance of GSM, UMTS, LTE and 5G cellular radio access technologies. 3GPP work is coordinated by regional organizations representing Europe, USA, China, Korea, Japan and India. Since its start in 1998, 3GPP is publishing work items in **annual release cycles**. For example, Release 16 was issued in 2020. Each release contains a set of features providing functionality across GSM, UMTS, LTE and 5G and making sure that these technologies will coexist and interoperate.

3GPP cellular network standards in combination with governmental spectrum auctions are the foundation of a well-established **business model for operators** who are offering services to users of mobile phones incl. smartphones. These operators are called MNOs (MNO = "Mobile Network Operator"). With more than 5 billion subscribers worldwide, this is an unparalleled success story and a win–win situation for MNOs as well as for network users. An industry organization called **GSM Association** (short: "GS-MA", [4]) represents the interests of worldwide MNO members—and besides 3GPP—a good starting point for potential users and other parties interested in cellular network technology.

Fundamentally, all 3GPP cellular networks are based on a large number of adjacent signal areas called **cells**. These cells join or overlap each other. Within each cell you will find a **base station** or **cell tower** which sends and receives data. Base stations are access points allowing user devices to connect to the MNO network. For IoT devices, a connected base station is acting as the entry point for Internet communication, i.e. to transmit IoT payload data or receive control information (see Fig. 1.5). Each cell is covering a certain area providing wireless Internet access to user devices within reach.

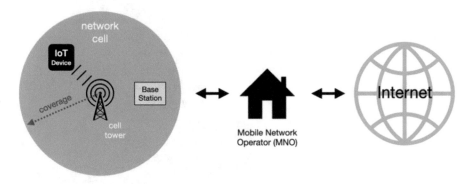

Fig. 1.5 Cellular internet access

A cell tower is a platform where antennas and other hardware are being mounted. Physically, this could be a dedicated mast or a building. In general, cell range (or cell size) depends on used antenna and applied output power (which might be restricted/limited by regional law). In rural areas, a single network cell will be able to cover an area of up to 35 km. By nature, cell **coverage** area will be small in urban areas because high-frequency radio waves cannot easily pass building walls easily. On the other hand, high population in urban areas means a lot of users might request service at the same time. But each cell can handle a limited number of user connections only. This means that a new connection request would be rejected by a cell if its **capacity** is exhausted. This gap will be compensated by additional cells within reach of a user terminal, i.e. In high-demand areas multiple cells with overlapping cell ranges areas will be provided.

Originally, cellular networks have not been designed to meet **IoT requirements** right from the beginning. In fact, some early IoT-usable features have been offered by cellular networks since 2nd generation, e.g. SMS. GPRS (General Packet Radio Service) for data transmission embedded in cellular radio network GSM was opened back in year 2000. But since LTE enhancements for machine-to-machine (M2M) communication have been included in Release 13 (2016), IoT connectivity became an integral part of cellular technology. Dedicated IoT network technologies NB-IoT and LTE-M have been specified a couple of years ago, and worldwide network deployment is growing continuously (see latest NB-IoT and LTE-M coverage maps [5] published by GSMA).

The capacity requirements target in 3GPP TR43820 [6] for Release 13 has been set to 40 devices per household which corresponds to 52.500 devices per cell or 60.680 devices per km^2. In combination with a high density of small cells in urban areas state-of-the-art cellular LPWAN network implementations will enable massive connectivity. In fact, a mobile phone or IoT device usually connects to the closest available **base station**, if not congested by other user devices. As an alternative, it may connect to the next available channel offered by another cell within reach. Figure 1.6 illustrates a typical scenario for

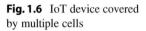

Fig. 1.6 IoT device covered
by multiple cells

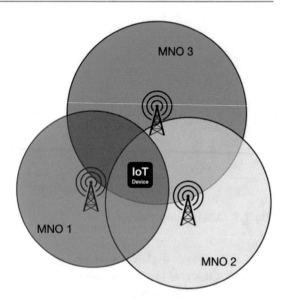

an IoT device in need for a data connection to a cellular network. In this case, the IoT
location in within reach of three different cells—operated by different MNOs.

Access to a cell will be granted only if the requesting user device has a valid business agreement with the owning MNO is place, a so-called subscription plan. For each
cellular IoT project, selection of a suitable connectivity partner is a key ingredient for a
successful rollout because all locations of an IoT device must be covered. GSMA coverage maps [5] and MNO information will be a good starting point. In addition, so-called
"virtual" network operators (MVNOs) are offering cellular network services which are
based on business agreements with these network infrastructure owners (MNOs). Consequently, a MVNO might offer multiple access points—operated by different MNOs—for
IoT devices at its particular deployment location. This strategy extends the usable overall
network coverage area. As different MNOs use different configurations for their network,
for an IoT deployment it is useful to check all of them, esp. if application-specific resp.
optimized network parameters are required, e.g., to achieve lowest power consumption
for battery-powered IoT devices (for more information how to analyze and adjust cellular
LPWAN network parameters refer to [1]).

As explained, use of cellular IoT networks is not for free, i.e., usage fees and subscription plans apply. At first glance, they look expensive. Wi-Fi and other wireless networks
in unlicensed spectrum appear as complimentary and a good deal. But this is only part of
the story. More important for a reasonable competitive comparison is the so-called **"total
cost of ownership (TCO)"** criteria which is calculated in completely different way for
licensed vs. unlicensed networks. A major differentiator is based on the fact that cellular network operators are independent service providers. They own the complete network
infrastructure and will provide IoT connectivity essentially everywhere as a service. In

contrast, an unlicensed connectivity infrastructure will be installed case-by-case at locations where IoT connectivity is needed for specific purpose, e.g., in office building or at home. These installations are usually owned by the user who will have to provide network access points, user management, security, sufficient network capacity, etc. Dedicated maintenance staff must be hired to ensure service delivery, software updates, support, etc. Related investments and operation expenses are impacting TCO and should be calculated for a realistic and fair comparison of licensed vs. unlicensed connectivity solutions.

Another good reason to select a cellular IoT network is **scalability** which is a critical requirement for massive IoT deployment scenarios. This means that used network must be able to handle a growing number of users or changing coverage conditions. For example, some countries have initiated **national rollouts of smart meters** in order to track power consumption in households. For this purpose, country-wide and robust coverage network coverage is mandatory because IoT devices might be located virtually everywhere and require reliable connectivity even under worst conditions, e.g., if installed in underground basements. For this kind of large-scale IoT projects benefits of a cellular network technology are unquestioned.

Another major benefit of a cellular IoT solution is that used network infrastructure is **always up-to-date** and future-proof. Cellular networks are based on **global** standards with very large industry support including manufacturers, network operators and service providers. **Long-term support and reliable service delivery** is quasi guaranteed because of billions of users all over the world are using it.

In general, each IoT project is aiming at different environments and facing different conditions at target locations. One of the key prerequisites for a successful IoT concept is a clear understanding of **requirements for deployment** of IoT devices. How many will be installed and where? In general, cellular network devices are easier to deploy because connectivity depends on network coverage only: if the IoT device is within reach of a network cell, Internet connectivity will work for this device. At first sight and for example, using an existing Wi-Fi network free-of-charge at point of deployment looks like a reasonable approach. But will it work, will local network coverage reach all locations and provided capacity guarantee for reliable service of all IoT devices? In fact, an existing network installation can help, if well-known and under control. In all other cases or in case of uncertainty resp. mission-critical IoT applications, use of cellular IoT networks should be considered—at least as a fallback option. If cellular network technology is used, no specific knowledge nor re-configuration of local IT network infrastructure (e.g., routers) will be necessary. This might be beneficial because installation of cellular IoT devices is **independent from existing local IT landscape**, if available.

Cellular IoT connectivity has two types: narrowband IoT (NB-IoT) and LTE-M. Although both are based on LTE cellular standard, there is one big difference between the two. That NB-IoT has a smaller bandwidth than LTE-M, and thus offers a lower transmission power. In fact, its bandwidth is $10 \times$ smaller than that of Cat-M1. In general, cellular IoT offers excellent reliability and qualifies for mission critical applications. In addition, device operational lifetimes are longer as compared to unlicensed LPWAN. But when it comes to choosing between one of them, **NB-IoT** is often preferred because it addresses typical IoT applications at **lower cost vs. LTE-M**. Typical IoT applications are **infrequently sending a small amount of data, e.g., meters or detectors** (see [1]). On the other hand, LTE-M offers mobility support and voice data transfer which is useful for transportation and tracking applications.

Note: For in-depth technical information about NB-IoT and LTE-M cellular network technologies see [7].

1.5   IoT Security Scope

An IoT ecosystem is based on several key elements (recall Sect. 1.2): IoT devices (incl. peripherals), a network and a central application server. All of these elements should contribute to the overall protection of the IoT application. Each of them is vulnerable and a potential entry point for data attacks or subject to accidental misuse. IoT security techniques are applied to ensure safe operation of all elements and take effect on several levels—by mechanical protection as well as by electronics and embedded software measures. The overall requirements for quality and strength of implemented security countermeasures of the IoT ecosystem depend on target application (see Sect. 1.5.2). In general, applied IoT security methods can minimize risks of data confidentiality/integrity breaches. Finally, IoT security is a "sales enabler" and ensures delivery of reliable and trustworthy IoT applications to the market.

The good news for IoT project owners is that, for a cellular IoT network, a specified level of security will be managed and guaranteed by professional network service providers (see Sect. 1.4). Same applies to the central IoT application if used server platform is managed by an external professional service provider (we will take a closer look at this topic in Sect. 4.3.2). As a result, IoT project owners will "just" have to select network and cloud partners carefully, then **focus on secure IoT device design** in order to meet required security objectives for the overall IoT ecosystem. Very often, IoT devices will have to work in public or unattended areas which are not controlled by the IoT service provider. Consequently, IoT devices are exposed to potential attackers and more vulnerable as, for example, an IoT server located in a safe environment. This is an extra challenge and must be considered carefully during specification of a secure IoT device.

Main design objective of an IoT device is (see Sect. 1.3) is to transmit collected raw or pre-processed local measurement data to the central server, listen to control commands from the central application and manage local execution. On top of this, a secure IoT frontend device design will have to ensure

1. protection against physical intrusion

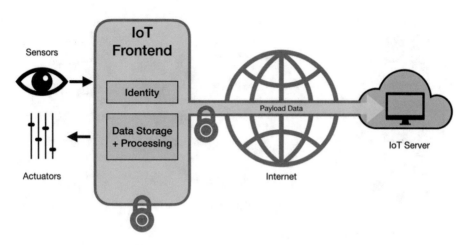

Fig. 1.7 IoT frontend security

2. proper execution of the embedded IoT application
3. reliable operation of connected peripherals (sensor, actuators)
4. establishment of a secure end-to-end communication channel to the application server.

In fact, the IoT device executes sensitive embedded software and processes sensitive IoT data which is finally transmitted to a safe place for further analysis. All credentials and cryptographic key material used for secure communication with the server are stored inside the IoT device.

Figure 1.7 illustrates the objective of security measures to be applied to the IoT device including communication channel to the server. The device itself is protected against physical intrusion or software attacks or attempts to exploit data leaks via side-channel attacks or fault injections. Finally, the device will act as a secure frontend device which is covering the complete installation of a local IoT application.

1.5.1 Communication Channel

By nature, IoT devices are connected to the Internet. This connectivity approach offers ultimate flexibility and allows deployment of IoT devices virtually everywhere world-wide. For IoT applications the network is used to transmit local IoT data to the server as well as to receive control commands (incl. firmware updates) by the IoT device. In order to secure IoT data transmission, security methods will have to makes sure to pro-tect integrity, confidentiality, and authenticity of sensitive payload data. Typically, data encryption and digital signatures are being used for this purpose (Sect. 3.1 will explain the basics). The idea is to put end-to-end encryption in place so that only the intended

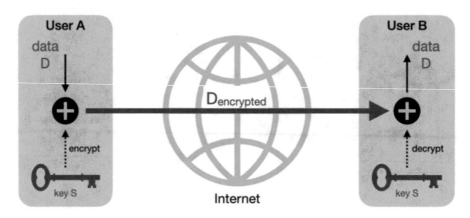

Fig. 1.8 End-to-end encryption principle

recipient of a message can read it. Figure 1.8 illustrates the functional principle of an end-to-end encryption of payload data sent by user A to user B.

In this simplified example, both communication partners are using a shared cryptographic key S for data encryption and decryption, i.e., a symmetric algorithm is being used. A previously agreed algorithm will be used by user A in conjunction with key S to encrypt data package D and convert into an unreadable cryptogram $D_{encrypted}$. Key S is securely stored in user device A, e.g., an IoT device. On its way through the Internet the encrypted data package $D_{encrypted}$ is perfectly safe if used algorithm and key length are strong enough. This method works independently from selected network technology, so that even the network operator or governmental authorities (!) can understand its contents. Only user B can decrypt it with key S and return it into cleartext data package D.

On top of data encryption, the same method can easily be extended in order to secure the integrity of transmitted data packages and authenticity of communication partners as well. For this extended solution, term **end-to-end security** is being used (instead of end-to-end encryption). This approach can also be used for secure firmware updates of IoT devices, if required.

For the IoT project owner, the end-to-end-encryption method offers ultimate data security and flexibility but requires protected local storage for cryptographic key(s) inside the IoT device. Required strength of protection depends on application-specific needs and overall security goals. Various soft-and hardware options are available as standard building blocks for IoT device designs, see Chap. 4.

1.5.2 End Devices

Security for IoT devices will not be available free-of-charge. Quality and strength of protection measures will require dedicated know-how, extra development efforts and maybe even additional components. Inevitably, corresponding cost will increase product sales price. But for most commercial IoT projects, return on investment (ROI) will justify extra cost. Related decisions are driven by target market requirements resp. competition, but other aspects apply. Some first thoughts to be considered for IoT security requirements are outlined in Fig. 1.9.

Key objective is TRUST. As explained in Sect. 1.1, IoT applications very often compete against use cases requiring human interaction, i.e., use of IoT techniques will lead to reorganization of processes and replace some jobs. Consequently, these social aspects will cause objections raised by affected people. But for sure, an unquestionable reliability and trustworthiness of IoT automation applications will help to silence critical comments.

In general, a secured operation improves overall product quality which has a positive impact on customer assessment. In fact, **IoT security is a selling point.** On the other hand, product failure can cause indirect commercial damage to a service provider, a manufacturer, etc. From a sales/marketing perspective, trusted operation of offered IoT products will help building a "zero-failure company" image which will influence customer purchase decisions and commercial success. A strong corporate image convinces customers—independently from a fact-based review of a company's product and services. "Trust sells". This fundamental truth applies to all IoT products.

Of course, built-in security is expected for some IoT products address traditional security applications like surveillance or asset tracking. In this case, IoT products will have to meet application-specific security requirements anyway, pre-defined either by customers or by legal standards (see Sect. 2.9). IoT products might even have to undergo security evaluations, and an associated security certificate might be prerequisite for qualification as a supplier, e.g., for public tenders (see Sect. 2.10).

In any case, for each IoT project a comprehensive **risk analysis** might lead to additional IoT security requirements. Following chapters "Threat Analysis (STRIDE)", "Vulnerability Analysis—Penetration Testing" and "Attacks" will take a close look at this important topic and at which level an appropriate set of countermeasures should be implemented. Who would be interested to misuse or tamper the IoT application? Which part of IoT data could be subject of an attack? What are potential impacts or quantifiable damages or financial loss. As a starting point, it will be necessary to understand if a deployed IoT device will be vulnerable to physical manipulation or intrusion by unauthorized people. Locations which are controlled by the IoT operator do not allow physical tamper attempts, but a an unattended IoT location in a public environment will provide additional options to the hacker (see Sect. 4.1). But even if deployed in a safe environment, an IoT device will always be vulnerable via injected malware or network attacks. Chapters 3 and 4 will

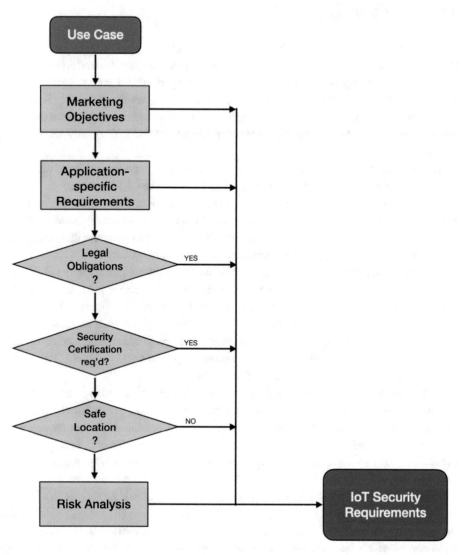

Fig. 1.9 1st-level considerations for IoT security

explain in detail which countermeasures can be used and which standard building blocks are available for design engineers.

References

1. Heins, K. (2022). *NB-IoT Use Cases and Devices—Design Guide.* Springer. ISBN: 978303084973. https://link.springer.com/book/9783030849726.
2. 3GPP—Third Generation Partnership Project. www.3gpp.org.
3. Withers, D. (1999). *Radio Spectrum Management: Management of the Spectrum and Regulation of Radio Services.* Institution of Engineering and Technology.
4. GSMA—GSM Association. (2021). Retrieved Dec 23, 2021, from www.gsma.org.
5. GSMA—GSM Association. (2021). Mobile IoT Deployment Map. Retrieved Dec 9, 2021, from www.gsma.orghttps://www.gsma.com/iot/deployment-map/.
6. 3GPP TR45820 (2015). Cellular system support for ultra-low complexity and low throughput Internet of Things (CIoT). 3GPP Technical report.
7. Liberg, O., Sundberg, M., Wang, E., Bergman, J., Sachs, J., & Wikström, G. (2019, Nov 28). *Cellular Internet of Things.* Academic.

Challenges and Objectives

2

2.1 Why IoT Security Concerns All of Us

IoT applications are entering our day-to-day work lives and homes. Due to the enormous number of deployed IoT devices (12.3 billion connected devices in 2021 [1]), these IT installations are becoming an increasingly attractive target for cyberattacks, particularly aiming at smart home accessories. According to cybersecurity expert Kaspersky, attacks addressing IoT devices have doubled over the past year [2]. In fact, any device which connected to the Internet is a potential entry point for attacker to generate damage on various levels, to steal data, manipulate operation or gain unauthorized access to services.

Besides potential damages caused by **IoT cyberattacks**, the societal impact (see Sect. 1.1) of IoT applications has been widely recognized. IoT is raising privacy concerns which are under permanent public observation. IoT is creating jobs, but also replaces others at the same time. This evolution is being eyed suspiciously by affected people and trade unions.

On top of this, manufacturers of IoT equipment and IoT service providers might be interested to generate a positive brand image and reputation as a "trusted company" in order to influence customer decisions from a top-level non-product-based perspective (see also Sect. 1.5.2). There are many good reasons why IoT service providers should take extra care about public perception, and why they will have to deliver unquestionable performance at any time. Of course, offered IoT service expects to receive trustworthy IoT data which is 100% reliable. Otherwise, usability and overall business value of an IoT application will suffer. This is a universal requirement for all IoT applications, but for some of them (e.g., mission-critical operations or management of confidential payload data) it is obvious an extra level of IoT security must be implemented by design. **Security matters to all IoT devices**, and typical objectives for manufacturers of trusted IoT devices are to.

© The Author(s), under exclusive license to Springer Nature Switzerland AG 2022

K. Heins, *Trusted Cellular IoT Devices*, Synthesis Lectures on Engineering, Science, and Technology, https://doi.org/10.1007/978-3-031-19663-8_2

- ensure **reliable and trustworthy operation**
- offer perfect data confidentiality, integrity, and authenticity
- protect identities and ensure of privacy of communication partners
- protect IoT ecosystems against fraud and counterfeiting attempts

On our journey to meet these objectives we will take a close look at requirements for trusted IoT devices and how to implement them. Focus is on successful prevention of cyberattacks. This chapter explains which IoT use cases are at risk and how to assess them.

Out of scope are risks leading to failures because devices were (accidentally) running under excessive environmental conditions, i.e., beyond specified maximum operation parameters like temperature, supply voltage, etc. This kind of risk applies to every embedded system which is supposed to work reliably and accurately. This requirement is called **safety** and usually not connected to cybersecurity requirements. But for vulnerability analysis and carefully targeted attacks against IoT devices, excessive operating conditions are applied by purpose in order to change its behavior. Section 4.1.2 will provide further information.

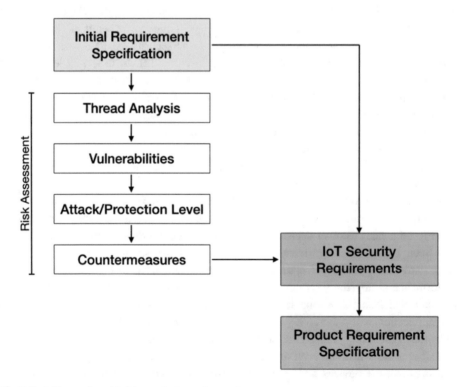

Fig. 2.1 IoT security added to product requirements

Accidental misuse of an IoT device is not covered by risk analysis—which usually does not apply anyway because a typical IoT does not offer physical user input options (e.g., buttons). In general, all risks against an unwanted change of a product are a problem. This also applies to its look-and-feel (brand protection), but not covered by this book.

Very often, threads against trustability are hidden or not obvious at first glance, so we will have to explain typical IoT risks and identify applicable threads first. Then, we will look at device vulnerabilities and attack scenarios. By nature, IoT devices are particularly exposed to risks because usually they are working in unattended and remote locations. This means that on top of **application-specific requirements for a IoT device**, risk assessment will lead to a set of **additional security requirements** which are contributing to the final product requirement specification. Figure 2.1 is illustrating this process.

2.2 Critical Use Cases

In general, an IoT application is sensing a remote location, analyzes retrieved data and takes an appropriate action. List of target IoT use cases is endless and covers many aspects of our life. Sometimes use case scenarios are being named according to their target markets, e.g., "Smart Factory", "Smart Home", "Wearables", "Connected Cars", "eHealth", "Smart City", "Smart Grid", etc. According to market researcher IoT Analytics [3], these are the top IoT applications areas and global market share:

1. Manufacturing / Industrial—22%
2. Transportation / Mobility—15%
3. Energy—14%
4. Retail—12%
5. Cities—12%
6. Healthcare—9%
7. Supply Chain—7%
8. Agriculture—4%
9. Buildings—3%
10. Other—3%

IoT application areas are different, but functional similarities apply. For example, predictive maintenance is an attractive use case for industrial production sites and as well for transportation vehicles or for smart grids. Figure 2.2 is illustrating typical IoT application categories, functional areas and target objects or environments.

IoT applications are addressing different business cases and each of them is following a specific set of technical, legal, and commercial rules. From a risk analysis point of view, the overview shown in Fig. 2.2 (see also [1], pages 8 ff.) might give a first idea which IoT applications are more critical than others. In fact, many IoT use cases do

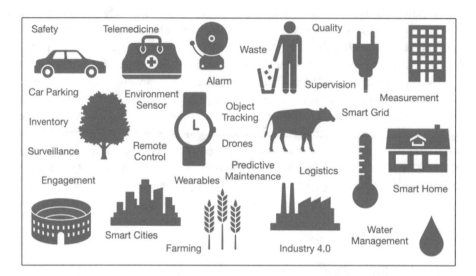

Fig. 2.2 IoT use case areas

not require any in-depth risk assessment because they are obviously not critical at all. For example, an **environmental sensor** delivering a wrong humidity value for a remote location might be problem but will not cause big financial loss. Same applies to public waste collection based on IoT-enabled litter bins: missing one of them during a daily collection tour will not kick the operator out of business. Other examples for uncritical apps are used to gather information rather than interaction with the target environment, e.g., for engagement tracking of visitors of an event.

On the other hand, for a critical case a cyberattack can cause damage to the **value proposition** of an IoT service—by changing the intended operation of the associated IoT ecosystem. This value proposition is closely related to the specific business model of the offered IoT service, i.e., which parties will suffer in case of failure. But from a technical point of view, quality of an IoT service is based on criteria like reliability, accuracy, data quality, performance, etc. And it depends on bullet-proof IoT functions and parameters delivered by a remote embedded system. For each IoT ecosystem, the owner will have to identify all **vulnerable design components** (embedded software, credentials, components, storage, data paths, busses, peripherals, etc.) and protect them against unintended change.

Table 2.1 presents a couple of simplified and idealized IoT use cases in combination with an associated business case—in order to prepare for next chapters which will take a closer look at threads, vulnerabilities, and countermeasures. Based on taken assumption and for a selected sample potential failure scenario, the associated possible cause of this failure is being named in Table 2.1.

These cases are simplified and idealized, but they are outlining various impacts and different IoT device design elements potentially causing damage in case of tampering.

Table 2.1 Sample IoT failures and causes

	Use case description	Failure scenario (example)	Impact, damage	Possible cause of failure (design element)
Temperature sensor	Travel agency providing free-of charge customer information	Device transmits wrong temperature value	company image might suffer	Broken connection to sensor
Baby phone	monitors remote noise activity	Activation by unauthorized person	Violation of customer privacy, legal implication (GDPR)	User identity verification function failed (embedded software)
Car tracker	provides actual location to the owner	Device transmits wrong position	financial loss	Embedded GPS function failed (antenna problem)
Flood detector	Alerts about water intrusion or leakage	Device does not submit alert	Water damage	Low battery (power supply)
Fire sprinkler	Remote controlled fire detector and water sprinkler	Device alerts server and receives activation command, but does not activate water dispense	Fire disaster financial loss, legal actions, liability case	Actuator control function failed (software problem)
Pay-TV set-top-box	Access control to video content	User consumes TV content free-of-charge	Financial loss of service provider	Device duplication (cloning)
Medical infusion pump	Remote controlled medication	Device injects wrong dosage	Patient health	Software bug (application program)
Room access control	Video surveillance system (face recognition) to submit an alert if an unknown visitor has been detected	Device does not submit alert	Loss of corporate assets, criminal activity	Device not reachable (sabotage)

Typically, these design elements are either functions or data delivered by sensors resp. processed or stored inside the IoT device. For each sample IoT device in Table 2.1, only a single failure has been selected as an example, but—in real life—multiple potential failures will be valid for each use case. Risk analysis will have to consider all of them and will finally lead to reasonable set of security measures—focusing on IoT device design (as explained earlier in Sect. 1.5). IoT applications usually are excluding human interaction and avoiding manual processes which are recognized as typical weak links in the security chain. Most threads are aiming at the IoT device which is the most attractive target for attacks.

2.3   Threat Analysis (STRIDE)

For a successful IoT service, owners will have to identify potential risks and put appropriate countermeasures in place. In most cases, IoT devices are the weakest link in the IoT security chain. Consequently, **risk assessment** is a key initial step for development projects aiming at a tailored IoT device associated with planned IoT service.

As a starting point for risk assessment, the business owner will have to identify potential threats and damages. Who might be interested to misuse resp. change specified behavior of your IoT product? What can go wrong? How likely is it? What are the consequences? What kind of damage would be caused by a failure of the application; can you quantify the amount of damage? Answers to this kind of questions need to be provided by the risk assessment process. Purpose of risk assessment is to **identify, estimate and prioritize risks**. As outlined before, there are many good reasons to put IoT security measures in place (recall Fig. 1.9: 1st-level Considerations for IoT security), but risk assessment outcome is critical as it provides the foundation for the identified risks to be mitigated. This is a serious topic because most **security breaches are not revokable, and most design elements of an IoT device are fixed** until product end-of-lifetime. As such, the product owner will have to take care as early as possible and incorporate requirements for countermeasures against identified risks into device specification, i.e., before starting the development process. Threat analysis should be a standard task in each IoT project plan.

Practically and in an effort to **identify all potential threats**, it is recommended to invite various stakeholders to a related brainstorming session where each of them can share their view from different perspectives: product manager, sales, marketing, customers, users, etc. (see Fig. 2.3). This experience will provide extra confidence to the IoT product owner who will benefit from concerns he/she might not have considered before. Related stakeholder concerns will fuel the associated tread analysis as a starting point for the risk assessment process. Starting point of a thread analysis is the **target use case** to be addressed by the planned IoT application and the associated **value proposition of the IoT product/service**, i.e., how this new IoT product or service is supposed to generate profit (see Fig. 2.3).

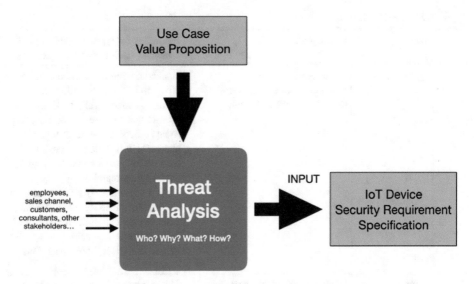

Fig. 2.3   Identification of potential threats

Several methods are available to support the risk assessment process. Some of them originally were aiming at threats against standard computer systems, but can be used at least as a starting point to analyze IoT vulnerabilities. One of them is the **STRIDE** model which is has been developed by Microsoft [4]. STRIDE is grouping threats into categories in order to formalize security concerns. STRIDE is a mnemonic and stands for these threat categories:

- Spoofing
- Tampering
- Repudiation
- Information disclosure
- Denial of service
- Elevation of privilege

Let's take a close look at these threats, explain each of them and how they address typical IoT usage scenarios. In addition, typical attack scenarios and potential damages are outlined in Table 2.2.

As a first step within the risk assessment process, the reviewer (e.g., the product owner) will first have to figure out for each STRIDE threat whether it applies to the characteristics of the assessed IoT application. Which of these threads is giving cause for concern?

Table 2.2 STRIDE threads overview

	Threat	Description	Examples for typical attack scenarios and/or potential damages
S	Spoofing	Malicious access to **authentication information** (held inside the IoT device)	Hacked credentials can be used to clone an IoT device ("identity theft"). A successful attack can be used to build a fake device claiming to be authentic. Based on this, a cheater will be able to access a billable IoT service without paying for it (e.g., streaming TV) On top of remote software attacks to read and disclose stored credential data, physical attempts may be used to intrude internal memory chips or to scan electrical interconnections inside the device
T	Tampering	Unauthorized **modification of data or code** (held inside the IoT device)	This method can be used to modify retrieved local IoT data and transmit fake IoT data instead. For example, an attacker can transmit wrong consumption data for billable goods (e.g., electricity) and reduce related payment. Related attacks typically will have to bypass authorization schemes or authenticate with a suitable user profile (e.g., administrator rights) allowing to update data or replace executable code
R	Repudiation	**Denial of ownership** of a closed agreement (e.g., contract, order, request)	Damage occurs to an IoT service provider, for example, if a customer disputes that he has submitted a specific online request for a payable service, and the contractor cannot proof a legally binding receipt of corresponding request. This thread is closely related to threads "S" and "T" because an attacker can use a spoofed customer identity for the request or related records have been modified (stored inside the IoT device)

(continued)

Table 2.2 (continued)

	Threat	Description	Examples for typical attack scenarios and/or potential damages
I	Information disclosure	**Exposure of sensitive information** to unauthorized individuals	Sensitive IoT data (e.g., measurement results) as well as device identity credentials are subject of potential attacks. Well-known "man-in-the-middle attack" (see Fig. 2.11) can be used by an intruder to read confidential data in transit between an IoT devices and the associated IoT server. Damage will particularly allow to fake submitted IoT data using a wrong device identity. Leakage of personal data might be a breach of data privacy laws (see Sect. 2.9)
D	Denial of Service (DoS)	**Disruption of service** of a target IoT device, i.e., it will be not available for its intended use any more	This thread aims at weakening an IoT business model or involved companies. Well-known "ransomware" attacks are inserting malicious software in an effort to extort the IoT business owner. For ransomware attacks, prior tampering (see thread category "T") of IoT code will be required. But simple physical attacks against an IoT device will achieve same result if deployment allows an attacker to access the device location
E	Elevation of privilege	An unprivileged user gains **privileged access** and thereby has sufficient access to compromise or destroy the entire system	This thread applies only to IoT devices which are built for different user profiles, e.g., which are based on a multi-user embedded operating system. For typical IoT devices with a single type of authorized user (with full administrator permission), thread category "S" applies

2.4 Vulnerability Analysis—Penetration Testing

In an effort to obtain access to internal resources, attackers typically will have to use vulnerabilities of the network interface resp. its API (see Sect. 2.6). Whoever is aware of the IP address of the IoT device is able to submit commands which are supported by the

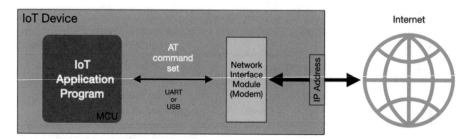

Fig. 2.4 IoT device internal versus external APIs

device. An attacker will have to explore which services have been implemented by the IoT device manufacturer to be available for network users, for example Telnet (Terminal Emulation) or FTP (File Transfer Protocol). In general, this kind of embedded services are provided by the used network interface module which is offering a versatile software stack (firmware) to be used by the IoT device designer for customization purposes. These firmware functions are available via device-internal AT-commands only (see Fig. 2.4), i.e., not from outside. For example, u-blox SARA-R5 modules allow to set up the IoT device as an FTP server (see [5]), but configuration details and access credentials will be disclosed to authorized users only.

In general, remote attackers could benefit from using AT-commands, but all modem chips are designed for control by a local host via UART (or USB) interface only. This means that an attacker would have to modify the embedded IoT application program (firmware) which is not accessible either. In fact, the network interface is a key component of every IoT device design because, besides its modem functionality, it offers a variety of administration functions, e.g., for file management including an update option for the device firmware. This function might have been enabled be the device owner, e.g., for field maintenance reasons. In this case, an attacker might be able to exploit it for injection of malicious code.

This example illustrates why remote tampering will be a difficult exercise for the attacker if no further information about the target IoT device is available. As a matter of fact, **physical access** to an IoT device will provide valuable insights to the attacker because to allows to open the device, take some time, and perform an in-depth **hardware analysis** in an unattended environment. In particular, the attacker can easily figure out, which components are being used and how they interconnect. Based on this information, the attacker can recreate the original design data (schematics) of the IoT device. This process is called "reverse engineering". Of course, the schematics itself does not disclose any sensitive information per se but delivers crucial data for further investigation. Once the attacker is aware of the schematics, internal data busses and interfaces to peripherals (e.g., to connected sensors) can be used for analysis and reverse engineering of the device configuration, embedded software, and stored credentials. For sure, this kind of analysis requires a certain level of engineering expertise and tools (at least "level-2" according to Table 2.3), but related literature is available to any interested party (for example [6-8, 13–15]) as a starting point.

For hardware analysis, an attacker will first have to obtain a **sample IoT target device**. Depending on the deployment scheme of a specific IoT service, this will be no major obstacle if these devices are sold as consumer products (e.g., a wearable medical device) or if installed in uncontrollable places anyway (e.g., an electricity meter or home automation devices). After removing the case, analysis starts with the identification of used key components, i.e., microcontroller, storage devices, sensors. For most designs, **standard products** will used. This means that specifications (datasheets) are public and will provide many useful details of the component, e.g., functions, registers, interfaces.

Assembly of components and metallization of a PCB (see Fig. 2.5) are reflecting the schematics of the IoT device but conversion of a given PCB layout into corresponding design data requires routing analysis (traces, vias, solder pads) of each single connection between components. This process is called **PCB reverse engineering**. Goal is to reconstruct the original design data, which is not illegal per se, related tutorials are public literature, e.g. [9]. This work does not require a level-2 expert yet, but some advanced de-soldering skills and tools are required, esp. for removal of large BGA packages (BGA = ball grid array) which are sometimes used for network interface modules or MCUs.

Fig. 2.5 Assembled PCB of a sample IoT device

On top of this, analysis of multi-layer PCBs requires intrusion or x-ray methods to follow traces located on buried (i.e., invisible) layers which are connected by vias.

As soon as components have been identified and all interconnections have been consolidated as design data (schematics), datasheets as well as further software analysis will disclose potential design vulnerabilities and weak points for attacks. Figure 2.6 illustrates the top-level block diagram of a typical IoT device.

Schematics are telling attackers which interfaces and busses are being used for communication between components. This is where all sensitive data traffic takes place, e.g., IoT sensor measurement data representing billable consumption. Or user credentials (passwords or cryptographic keys) which can be used by attacker for fake authentication or data decryption/encryption. For analysis purposes it makes sense to probe these lines and scan for valuable information. Figure 2.6 indicates these data paths (marked **red**) which are inputs and/or outputs of the central MCU. This MCU is hosting and executing the embedded application program of the IoT device and all relevant data incl. sensor and actuator commands. The MCU is also mastering network communication and controls related modem functions via AT commands, but—after an initial setup by the IoT application—many network transport layer and security functions are running autonomously on the modem subsystem (aka network interface module). **Tracing** device-internal **AT commands** and responses allows an attacker to figure out how the network interface has

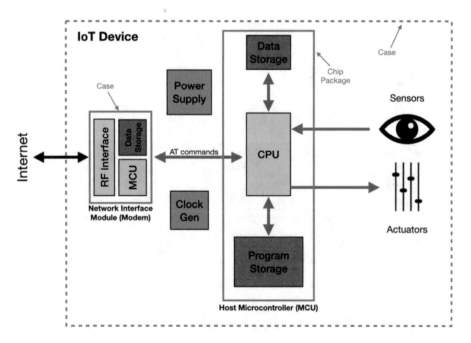

Fig. 2.6 Working points for penetration tests (marked RED)

been configured and which functions are available to external network users. e.g., for remote attacks.

I^2C is a popular serial bus which is often used as an interface to IoT peripherals. **Probing the I^2C bus** connection to an external sensor allows to analyze commands submitted by the MCU (master) and associated payload data returned by the sensor (slave). For this purpose, a digital oscilloscope can be used to measure and record electrical signals on SCL (I^2C clock) and SDA (I^2C data) traces. The upper part of Fig. 2.7 presents a sample oscilloscope screen shot with raw I^2C data to be used as an input for analysis. Prior hardware analysis has identified type and manufacturer of the sensor, datasheet provides details about implemented I^2C registers and commands. Finally, the PCB layout indicates that I^2C pins are used by the sensor. I^2C specification [10] explains how SCL transitions are used by the slave device to detect relevant data presented on the SDA line.

From inactive state, start of a command sequence is triggered whenever the SDA line switches from a high voltage level to a low voltage level *before* the SCL line switches from high to low.

Then, rising SCL edges are used to sample the actual level on SDA line, i.e., data on the SDA line must be stable during the HIGH period of the clock. First seven clocks are used to specify the I^2C register to be addressed (7-bits), bit 8 indicates that the MCU wants to read the contents of this register. Ninth SCL pulse allows the receiver to signal the

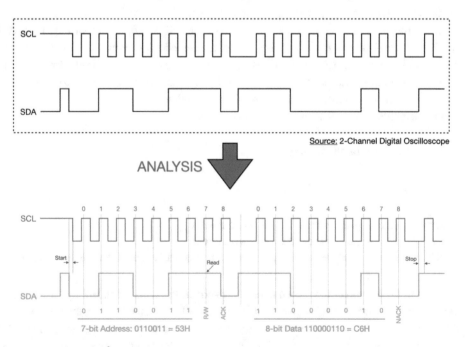

Fig. 2.7 Analysis of I^2C traces

transmitter that the byte was successfully received (ACK) and starts to return 8 bits of data from specified register. Typically, this data will be sensor payload data which has been requested by the IoT application, e.g., current temperature, a fluid level, consumption data. The sensor datasheet explains how to convert transmitted digital data into measurement values.

2.4.1 Software Analysis

The I^2C example is showing (Fig. 2.7), how **reverse-engineering of functions and internal command sequences** works—as a starting point to understand details of the IoT application. Of course, same approach also works for the IoT application program code itself—on its way from program memory to the MCU (see Fig. 2.6). Unfortunately, MCU programs are usually stored inside the MCU, on the same silicon. De-capping a chip and accessing an internal data bus is not impossible but requires a level-4 expert (refer to Table 2.3). Instead, from an attacker perspective, it will be much easier to use chip-internal facilities (e.g., a JTAG or SWD debug port) or MCU development tools to access this code. This exercise would probably require a level-2 attacker only, if the MCU offers this kind of entry door. In the end, penetration testing will figure out which configuration options and protection mechanisms have been applied for the IoT device. Based on these insights, the attacker will select the most promising and most efficient attack approach.

The ultimate goal of an attacker is to retrieve the full program code of the embedded IoT application—and to modify it. By nature, an embedded program is stored as binary code. But knowing the CPU core, disassembly tools are available to transform machine code into a human-readable representation of the instructions. Disassembly tools are even available online, see https://onlinedisassembler.com/odaweb/.

For example, short ARM machine code sequence.

```
0B00A0E31F10A0E3010080E0
```

converts into ARM assembler code.

```
MOV r0,#11 ; put value 11 in register 0
MOV r1,#31 ; put value 31 in register 0
ADD r0,r0,r1 ; r0 = r0 + r1. add r0 and r1 values and store in r0
```

Starting from the **reconstructed assembler code of the embedded IoT application**, an attacker can remove or change part of it (e.g., use different user identity, submit fake IoT data), recompile the modified code and replace the original program. For sure, most IoT device designs will prevent unauthorized people from dumping program code or replace it, but—on the other hand—most IoT devices will offer some kind of remote firmware update function. And there might be a way how to misuse or to circumvent it.

2.4.2 Side-Channel and Fault-Injection Attacks

Provided I^2C example was demonstrating how an attacker can obtain valuable functional details after some hardware reverse engineering of an IoT device. But besides physical intrusion there are other ways how to analyze behavior and internal resources. For example, more sophisticated side-channel attacks (SCA) or fault-injection (FI) techniques can be used to investigate implementation details of an embedded system and extract secret information like cryptographic keys or program code. It requires level-3 or level-4 expertise (refer to Sect. 2.6.2) to benefit from these kinds of attacks.

Physical fault-injection means to expose device components to extreme conditions beyond its specified limits, e.g., for ambient temperature or power supply voltage. Intention is to change system behavior and/or output by a carefully targeted fault injection. A successful attack can modify the program behavior in different ways such as corrupting program state, corrupting memory content, stopping process execution ("stuck-at fault"), skipping instruction, modifying conditional jump, or providing unauthorized access. Typical threats involve tampering with clock (freezing or glitch) and supply power (under/over voltage or glitch).

For example, a precisely timed clock glitch can be used to skip execution of a specific CPU instruction, see Fig. 2.8. This way an attacker can bypass unwanted firmware operations like updating a memory location with critical data. Or execution of implemented security checks can be avoided, e.g., for verification of the actual user identity or for sensing actual environmental conditions. Many fault injection attacks are non-invasive,

Fig. 2.8 FI causing corrupted program flow

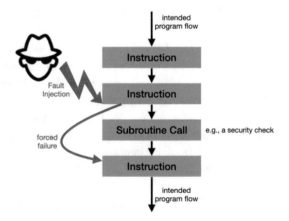

i.e., they leave no marks and are not evident when finished. For more information see [11] and [12].

Side-channel attacks exploit information leakages of the system. Even the system implementation allows authorized access only, a "side channel" might offer valuable insights to internal processes and used data. The leakages can be related to timing, power, electromagnetic signals, sound, light, etc. SCA is a non-invasive and passive attack. For more information see [13, 14]. For example, cache attacks might allow to retrieve sensitive information from a CPU allowing multi-user operation.

But for cryptographic IoT devices, SCA is a popular approach to retrieve keys which are stored and executed internally. Particularly, this applies in case a secure algorithm has been put in place, and brute-force attacks are not feasible due to large key space resp. long keys used. For this purpose, **power analysis attacks** can be used which are based on the fact that CPU power consumption depends on instructions and data being processed. Fundamentally, this is due to physical effects of how digital devices are built, i.e., changing the state of an internal digital line requires energy, i.e., power consumption slightly increases. These variations can be observed where power is being applied to the device, i.e., its power supply pin. A standard oscilloscope can be used to measure power consumption of the device. On the other hand, it is public information how standard cryptographic algorithms like AES work, i.e., the sequence and structure of computational tasks are being performed during each encryption/decryption session (see [15]). Based on this knowledge, it is possible to modify known values (e.g., plaintext for encryption) and compare/analyze sampled power traces allowing to select best guesses for used AES key, narrow down candidates, etc. until the AES key has been identified.

Section 4.1 will explain how to protect IoT devices against different kind of intrusion techniques.

2.5   Security Items for IoT Devices

When designing an IoT device, security measures can address most sophisticated attacks with highest level of assurance, see Sect. 2.10, but also countermeasures against level-1 attacks at zero additional cost are available. Figure 2.9 illustrates an overview of key IoT security items requiring protection (labeled "1" to "11") and associated indication of required investment (labeled "$" to "$$$"). **Security items no. 1 to 4** are referring to **basic security design aspects** to be considered in any case—even if no specific threads have been identified. **Security items no. 5 to 8** are referring to more advanced countermeasures requiring extra design efforts and expertise.

- **Security item no. 1 ("Initial Device Configuration").** Many off-the-shelf MCUs or network interface modules are delivered with inappropriate default configurations. As a general rule for a secure initial configuration, only essential network features or functions should be enabled. In order to minimize potential vulnerabilities, designers should disable network services like FTP or Telnet, if not needed. Embedded applications should be stored inside the MCU chip, not in external memory. MCU configuration should ensure that memory content is one-time-programmable and not readable afterwards. MCU debug ports and other facilities allowing MCU reconfiguration or access to internal resources or interfaces should be disabled. In order to avoid tampering in the field, initial configuration setup should be fixed permanently before the IoT device leaves production. For further information how to select a suitable MCU refer to Sect. 4.3.3.

- **Security item no. 2 ("Device Authentication").** IoT devices usually work at remote locations in unattended environments. To distinguish them, each device needs a **unique and trusted identity**. This identity can also be used for secure communication, i.e., for device authentication and key exchange in order to set up an encrypted data channel. Same mechanism can also be used for user authentication if the device is firmly coupled with its owner. This is important for repudiation scenarios (see Sect. 2.3). Managing device identities might be handled during device production in a cost-efficient

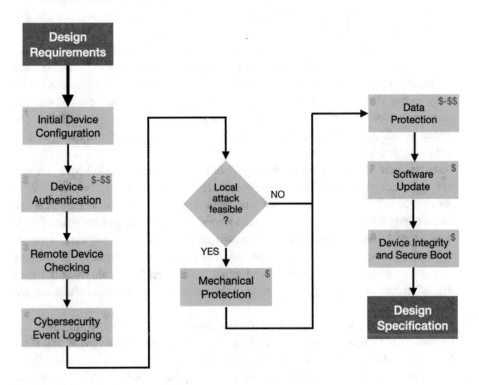

Fig. 2.9 IoT security items overview

way, but might be challenging if external services for registration, revocation, etc. are needed. This topic will be handled later in Sect. 3.3.

- **Security item no. 3 ("Remote Device Checking")** describes a mandatory function for remote IoT devices operated in uncontrolled environments. It allows to ping the devices and to ask them to return a sign-of-life and to confirm normal operation. Without this evidence, the IoT operator might misunderstand a "quiet" device esp. if this device is supposed to submit conditional alerts.
- **Security item no. 4 ("Cybersecurity Event Logging")** is referring to standard requirement for IoT devices to **detect and record relevant parameters** (incl. voltage, frequency, temperature variations) or other suspicious events or activities in internal non-volatile memory. Some dedicated additional sensors might be required. This log can be used later as tamper evidence or for investigation purposes by authorized users (see extra Sect. 4.1.3).
- **Security item no. 5 ("Mechanical Protection")** is valid for IoT deployment scenarios allowing local **physical attacks** during device operation (on top of logical attacks). Countermeasures are being introduced later in Sect. 4.1.
- **Security item no. 6 ("Data Protection")** is referring to the embedded IoT application including **program data (firmware), payload data and device identification**. These resources are reflecting the expected behavior of the IoT ecosystem and must be protected against modification. This is a fundamental requirement for all IoT ecosystems, regardless of applicable threat scenarios. Requirement includes data storage as well as internal data paths where sensitive data might get attacked. Besides data modification also **internal eavesdropping** is a problem because it allows identification of weak points (see Sect. 2.4) or cloning the identity of an IoT device. For **external communication**, e.g., with the IoT server, data protection is being prepared inside the device by adding a digital signature (message authentication) and encrypting data for transmission through the Internet.
- **Security item no. 7 ("Secure Software Update")**. A secure remote update function of the embedded software is another fundamental requirement, esp. for complex IoT devices. This is a **precautional measure, for example, to be used in case of a critical security breach** (e.g., tampered software) which would otherwise cause the IoT operator to initiate an emergency shutdown the IoT device. In this case, a remote software fix would allow to resume operation and avoid device replacement. Another example is a large-scale compromising attack against applied chain-of-trust for the IoT infrastructure requiring an exchange of all device identities and certificates (see Sect. 3.3)
- **Security item no. 8 ("Device Integrity and Secure Boot")**. This requirement goes beyond security item no. 6 ("Data Protection") and ensures that a device will return from power-down or reset to a trusted operational state. The idea is that the device should **not trust any code or peripherals** until it has been validated as being authentic and that its integrity is confirmed. See extra Sect. 4.2.3.

2.6   Attacks

Risk assessment process will have to determine which security requirements require attention. But for related decisions about strength of countermeasures to be implemented, we will also have to take a closer look at expected potential attacks, particularly at.

- different **types of attack** the target IoT ecosystem might be facing and
- applicable **skill level(s)** describing attacker expertise and budget.

2.6.1   Attack Type

For many IoT use cases, a local physical attack against a deployed IoT device will not be feasible because locations are attended or controlled somehow by the operator. In fact, most attacks are executed **remotely** as so-called **logical attacks** aiming at the IoT device in an effort to exploit internal data or to inject malicious software into the embedded IoT application program. For a remote attacker, the only way to reach internals of the IoT device is its Internet connection. Most IoT devices will be designed to allow interaction with authorized users, e.g., for remote control or update features. Attackers will try to exploit vulnerabilities of this network functions. This will be difficult for the remote attacker, if he/she is not aware of any further details of the target device. So, in an effort to identify suitable entry points, the attacker will try to obtain a sample device and analyze it, see Sect. 2.4. These insights will possibly allow to develop a dedicated remote approach how to access to the device configuration or embedded software.

Of course, physical access to a deployed IoT device in full operation offers many additional options for tampering. This attack scenario applies, for example, to smart meter installations in households. For a **local physical attack**, the attacker might be able to remove mechanical protection and access internal electronic components and data connections. For example, these access options allow to apply tailored spy attacks to sensitive data or credentials. Of course, you can also exchange a deployed device with a tampered device which has been prepared for this purpose. Device protection for this kind of physical attack requires **advanced hardware protection methods** we will introduce later (see Sect. 4.1).

An example for cyberattacks which are not addressing internals of the IoT device are used for "D" threads. In this case, attacks are not aiming at internal IoT application program code or IoT data, they just have sabotage in mind (maybe in combination with extortion or the service provider), i.e., disrupting the service. If the attacker has physical access, he/she can just simply destroy or unplug the IoT device. In the IT world, for the most common example of **Denial-of-Service (DoS)** attacks the attacker uses a cluster of hijacked computers. This group of computers is used to flood the target site with phony server requests, leaving no bandwidth for legitimate traffic. If the attacker is able to get hold of device IP addresses, same kind of attack can also address IoT installations—by taking connected IoT devices out of operation and jeopardize business. See Fig. 2.10.

Other attacks are not addressing the IoT device at all. For example, mentioned "I" attacks are address the connection channel between the IoT device and the central server (see Fig. 1.1). A so-called "man-in-the-middle attack" can be used by an intruder to eavesdrop a data transmission and extract sensitive information, see Fig. 2.11.

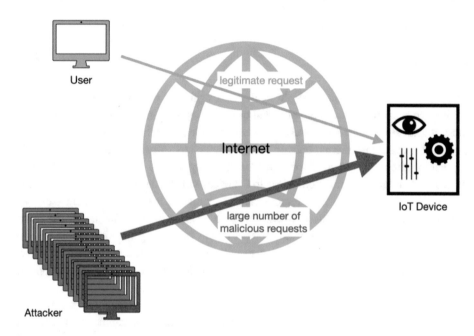

Fig. 2.10 IoT Denial-of-Service (DoS) attack

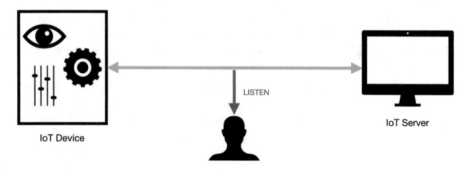

Fig. 2.11 Man-in-the-middle attack

2.6.2 Attacker Expertise

Thread analysis identifies potential attacks and impacts, but at the same time also provides an indication about potential financial loss in case of a successful attack attempt. This is, for example, claimed amount of ransom money to be paid in case of a DoS attack which is putting a valuable process out of operation. Or, in case of eavesdropping the credentials of an authorized user, the financial loss is equivalent to the value a billable online content (e.g., pay-tv) the attacker can consume for free. So, a key task is to quantify the commercial disadvantage caused by a successful attack. In fact, implementation of a decent level of IoT security is not available for free, but also cracking IoT protection requires investment on attacker side.

As a rule of thumb, identified value (of tampering an IoT device) will correspond with **expertise level** of a potential attacker resp. **budget** a potential attacker will be ready to invest. Higher level of expertise and larger budget will raise **attack quality**, enable more challenging attacks, and allows to circumvent more complex protection measures. In general, the hacker will have to penetrate the internals of a target IoT device and.

- analyze electronics hardware and embedded software ("reverse engineering"),
- identify vulnerabilities and most attractive business case how to exploit and monetize them,
- select the most promising entry door and develop a tailored attack approach,
- build an attack vehicle (software-only or in combination with an extra MCU device) and
- execute attack attempt(s).

For the attacker, the available budget will be in invested in lab equipment which is needed for hard- and software analysis (e.g., intrusion tools, oscilloscope, logic analyzer, de-soldering equipment), see next Sect. 2.4 for further details. But, in case of organized crime or a criminal group, this budget also reflects cost of additional expertise resp. time spent for target analysis as well as for development and execution of a series of attack attempts. Of course, combining skills of several experts allows to achieve results more quickly.

In an effort to categorize different attacker types, Table 2.3 describes typical characteristics and associated qualification.

Table 2.3 Attack levels

Level	Name	Skills
1	**Amateur**	Has a technical mindset and **good understanding** of electronics and low-level programming skills. He will not perform own complex analysis or reverse engineering, but he is able to follow tamper instructions, e.g., how to de-solder a memory chip and dump its contents and reprogram it with a new image. He can also create simple software attacks by connecting a pre-configured Raspberry Pi with a target IoT device via network or an external bus like I^2C
2	**Engineer**	Has a solid electronics **engineering background** and some programming skills, but he has no particular hacking experience. He learns quickly and is familiar to work in a lab and to use development tools incl. measurement equipment. He would certainly be able to hack an IoT device (with medium complexity and protection) but would have to spend time get familiar with specific needs and to achieve results
3	**Expert**	Similar to a level-2 attacker, he is a qualified engineer with solid electronics/programming skills and a certain level of criminal motivation. But on top of this technical background, he has **dedicated experience** how to hack embedded systems. He also has a basic understanding of crypto analytics. He has access to a **well-equipped lab** and will be able to perform advanced analysis of hard- and software (incl. side-channel attacks) and to create tailored local and remote attacks
4	**Ultimate**	Has **ultimate hacking capabilities** and a budget allowing to invest even further in order to prepare for most sophisticated attacks. If needed, a group of state-of-the-art level-3 experts and best possible lab equipment will be available for creation of most advanced side-channel attacks and intrusions even inside a chip. This attacker is driven by a high level of **criminal motivation**

2.7 Protection Level

Requirements explained in Sect. 2.5 will be incorporated into the design specification—at a certain level. Used term "Protection Level" reflects to the **strength of applied counter-measures** and will correspond to expected level of attacker skills (refer to last Sect. 2.6.2): 1-Amateur, 2-Engineer, 3-Expert, 4-Ultimate.

For level-1 countermeasures, no specific IoT security expertise will be required. But for level 2–4, dedicated security training will improve understanding of available options and how to implement them. This investment creates in-house IoT security expertise which will be valid beyond the actual IoT project. But as an alternative, also external consultancy can step in and provide support for security design aspects—at extra cost. Table 2.4 is outlining countermeasures for each protection level.

Table 2.4 Protection levels

Level	Name	Countermeasures
1	**Amateur**	• Focus on **proper configuration** of modem and host MCU components (security item no. 1 of section "Security Items for IoT Devices") • follow **secure design recommendations** and related application notes of component manufacturers • implement functions for **remote device checking** and event **logging functions** at no extra cost, i.e., without adding any further components, e.g., sensors (security items no. 3 and no. 4 of section "Security Items for IoT Devices")
2	**Engineer**	On top of level-1 countermeasures: • implement a cryptographic scheme for secure injection of a **device identity key during production** (security items no. 2 of section "Security Items for IoT Devices") • set up a **secure communication channel** to the IoT server (e.g., based on TLS protocol stack bundle of selected network interface module) • if applicable, implement a **basic protection against mechanical intrusion** (security items no. 5 of section "Security Items for IoT Devices"), e.g., security screws • implement a **basic data protection** method (security items no. 6 of section "Security Items for IoT Devices") allowing data storage and code execution inside a chip (i.e., on silicon level), e.g., by use of high-integrated solutions combining host MCU, memory and network interface in one single package (see Sect. 4.3.7) • add a **secure firmware update** function (security items no. 7 of section "Security Items for IoT Devices") in an effort to keep IoT devices alive even after a successful cyberattack • in order to gain extra confidence, a comprehensive expert review of a security concept and implementation is recommended. Some IoT security labs are offering independent consultancy and prototyping services plus certification. See extra Sect. 2.10
3	**Expert**	On top of level-1 and level-2 countermeasures: • if local attacks are applicable: extend device intrusion protection using additional **methods to detect and record suspicious events** (see Sect. 4.1.3). Goal is to achieve an advanced level of **tamper-evidence** • if software tampering is a top concern, besides measures for data and software protection, an advanced **device integrity and a secure boot** process is a further security option (see security item no. 8 of section "Security Items for IoT Devices") • implement an infrastructure for **controlled deployment of IoT devices and technical data** in an effort to exclude potential attackers from gaining technical insights or obtain a target IoT device for penetration testing (refer to Sect. 2.4). For this purpose, availability of technical documents as well as deployment or sales of IoT devices should be executed in a controlled/restricted way and to qualified and identified customers only, e.g., by putting a non-disclosure agreement (NDA) procedure in place. Sales channels and processes have to be managed accordingly. Another considerable approach is to provide an **IoT device only for rent**, not for sale
4	**Ultimate**	On top of level-1 up to level-3 countermeasures: • for ultimate strength of protection use of dedicated **tamper-proof electronic components** is required. For this purpose, CC-EAL4 certified integrated IoT solutions must be used. See extra Sect. 2.10 • Or so-called **secure elements** which are based on smartcard technology offering state-of-the-art protection against level-4 attacks (see Sect. 4.3.5)

2.8   Risk Assessment for Sample Use Cases

Figure 2.1 already mentioned major ingredients for risk assessment. For each IoT application, some STRIDE threads might apply, others not (recall Sect. 2.3). But even if a specific thread is causing concerns, it might not be rated as a critical risk requiring serious consideration for the IoT device design. Key question for each identified threat is: How likely will it occur? Do you expect a corresponding attack to occur after deployment of the IoT device during its active lifetime in the field? How likely will a potential damage materialize into a real financial problem or decrease of revenues due to caused loss of reputation of the IoT service provider? Consequently, a decent **criticality assessment** of an identified thread should justify implementation of a corresponding countermeasure.

Relevant criteria for criticality assessment are.

- potential **financial loss** or business value of caused damage for all applicable risk scenarios in case of a successful attack
- **attack quality**, criminal potential, budget (e.g., for equipment) and time required to prepare and perform a successful attack (refer to Sect. 2.6.2)
- **device location** (attended, public space? physical attack possible?)

In any case, security requirements apply (recall Sect. 2.5 in accordance with an appropriate protection level 1 to 4 (refer to Sect. 2.7). By nature, strong protection will be more expensive (expertise, development time, bill-of-material) compared with less secure implementations (e.g., by software only). Project owners will always have several technical options to choose from, but at least for non-governmental IoT deployments strength/cost of specified IoT security measures will have to match criticality of identified risks.

Final risk assessment is supposed to **specify a reasonable set of security measures** for each IoT project. Every IoT application has its own value proposition and follows a specific set of technical, legal, and commercial rules. As a consequence, risk assessment needs to be done case-by-case and separately for each IoT project. In order to illustrate the process, several reference IoT products have been used as examples. For this exercise and for each IoT application we

- describe use case and applied business model,
- motivation for attacks, likely attack scenarios, potential damage,
- a "risk scorecard" consolidating assessment inputs and results,
- security measures

Estimation of potential damage as a financial loss is not obvious for many IoT use cases. Typically, companies tend to underestimate damage. In fact, according to insurer Hiscox, the medium financial damage was not less than 57,000 USD per cyber incident in 2020 (see [16]). But it will be particularly difficult to quantify lack of revenues caused by a damaged company image. But "business value under normal conditions" is certainly a good starting point, if this value is available or can be derived somehow.

Validity as well as criticality level of each thread are contributing to the final risk assessment which consolidated in each risk scorecard. As a conclusion from our risk assessment journey, a list of suggested security measures is presented for each sample IoT product.

2.8.1 "Forest Sensor"

- **Description of use case and product:** IoT solution for monitoring health condition of trees. A battery-powered sensor device is mounted to a tree and collects vital and environmental parameters. This data is being consolidated and analyzed in a cloud and reports statistical and actionable information to forest owners. IoT product *Forest Sensor* consists of a device plus client software plus access to cloud service. It reports forest data on a regular base (e.g., once per day) in order to limit power consumption and increase battery lifetime.
- **Description of business model:** Business idea is to provide valuable information about forest health to the owner of a forest in order to optimize growth. Products are commercial products which are available to everybody for purchase via stores or online shops. Business owner is the product manufacturer and provider of associated service.
- **Motivation, attack scenarios, potential damage:** No financial value nor sensitive information is involved, so there is no high-level motivation to tamper or misuse *Forest Sensor* devices or service. But theft or vandalism might be a problem because devices are located in uncontrolled areas. In this case, service for affected device would discontinue, but financial loss reflects cost of replacement only.
- **IoT security concept (countermeasures):** Protection level-1 is sufficient (refer to Sect. 2.7). In particular, implementation of following measures is necessary:

  o Identities of damaged or stolen devices should be "blacklisted" by the manufacturer in order to exclude them from *Forest Sensor* services.
  o If a *Forest Sensor* does not report regularly or in case of doubt, users should be able to trigger the device to return a meaningful "sign-of-life" message indicating proper operation. If it does not reply, it will be gone (or battery is empty) and local action will be required.

Table 2.5 Risk scorecard "Forest Sensor"

STRIDE Thread (see Sect. 2.3)	Relevant ?	Attack Scenario (see text)	Potential damage	protection level (see Sects. 2.6.2 and 2.7)	Security items (see Sect. 2.5)	Countermeasures (see text)
S	yes	Theft	discontinued service, device cost	1	2	"Blacklist" device identity
T	no	n/a				
R	no	n/a				
I	no	n/a	n/a		n/a	n/a
D	yes	Disable activation	Value of room assets		3	Remote "sign-of-life" function
E	no	n/a	n/a			

2.8.2 "Room Guard"

- **Description of use case and product:** Purpose of this IoT device is to monitor human presence in a protected room, for example in a museum. An ultrasonic or infrared Time-of-Flight sensor is used to determine distances to objects and changes, if any. In case of detection of a moving object, the device will submit a configurable alert, e.g., an SMS message. *Room Guard* will be installed in environments which are controlled (physically or by other means) by dedicated security staff during business hours. *Room Guard* will take over after hours or at night or whenever no person should be present in this room. Activation/De-activation of *Room Guard* is done remotely via Internet by authorized staff.
- **Description of business model:** *Room Guard* products are sold directly or by authorized resellers. *Room Guard* is not offered as a low-cost consumer product. Instead, it is a professional B2B (business-to-business) product providing a specified level of security and reliability. This commitment is a value proposition which is attracting customers and implies that the product manufacturer has to take (limited) responsibility caused by tampered devices.
- **Motivation, attack scenarios, potential damage:** Motivation for unnoticed presence in a protected room depends on assets located in this particular room. Potential damage is the value of assets in danger, e.g., might get stolen or confidential information which might get disclosed to the attacker. Attacks are driven by criminal energy to disable *Room Guard* alert function in an effort to hide criminal activities in this room. No local attack will be possible, remote attacks will try to achieve one of the following results:

  o Activation of authorized staff has no effect, i.e., "Roam Guard" remains inactive
  o Deactivation of the device

Table 2.6   Risk scorecard "Room Guard"

STRIDE Thread (see Sect. 2.3)	Relevant?	Attack Scenario (see text)	Potential damage	protection level (see Sects. 2.6.2 and 2.7)	Security items (see Sects. 2.5)	Countermeasures (see text)
S	yes	submit fake deactivate command on behalf of authorized staff	value of room assets	**3/4**	1, 2	Strong user authentication
T	yes	modify device operation, change device identity	value of room assets		1, 3, 6	secure storage
R	yes	failure or side-effect of other attack	liability		1, 4	Command log
I	no	n/a	n/a		n/a	n/a
D	yes	disable activation	value of room assets		3	Non-IP activation method
E	no	n/a	n/a			No different user profiles

- Risk scorecard:
- **IoT security concept (countermeasures):** Protection level-3 (or even 4) will be required (refer to Sect. 2.7) to stop well-equipped criminals. In particular, implementation of following measures is necessary:

  o In order to eliminate spoofing attacks ("S") aiming at misuse of authorized staff identities, most important is a **strong user authentication** scheme. Tampering attacks ("T") aiming at modification of device software or data are less efficient and therefore less critical.
  o In order to avoid risks of vulnerability testing, sales to customers will be performed using a **restricted B2B sales channel** including a mandatory customer registration scheme. *Room Guard* products can be purchased by registered customers only.

o No local attacks will be possible either, so there is no strong need for hardware-assisted data storage. Proper software implementation will be sufficient.

o DoS attacks might be used to eliminate attempts of authorized staff to activate an *Room Guard* device. Therefore, a second **non-IP method** should be used **for activation/deactivation commands**, e.g., a SMS message.

o In case of doubt or on a regular base, device will be asked to return confirmation that it runs normally.

o In case of damage, an implemented function to maintain an **activity log file** can use for evidence.

2.8.3 "eMeter"

- **Description of use case and product:** *eMeter* is a **smart meter** product for electricity consumption in households. It is reflecting typical characteristics of real-world smart meter deployments which are ongoing in many parts of the world incl. China, India, Korea, Japan, France, Spain, UK, and Germany. According to market researcher GMI worldwide revenues already exceeded 16 billion USD in year 2021 [17]. Smart meters are used for automatic remote consumption monitoring—replacing conventional meters and associated need to retrieve this data manually. Smart meters are located at the energy point-of-deployment of the so-called **smart grid** which is the technical infrastructure to facilitate efficient energy supply, distribution, and consumption services. Our *eMeter* device is measuring, recording and reporting billable energy consumption.

- **Description of business model:** From a consumer point of view, business partners are companies managing supply of electricity at home. Usually, these electricity dealers are not owning powerplants themselves but act a sub-contracted sales channel on behalf of one or multiple energy supplier(s). The provider offers a customer package including a tariff (e.g., 30 cents per kWh) and an *eMeter* IoT device to be installed at place of consumption. The *eMeter* is reporting actual power consumption in real-time to the provider via Internet (i.e., a cellular network). This IoT data determines monthly billing to the customer which is submitted automatically. Main involved parties and communication channels are outlined in Fig. 2.12.

In fact, the *eMeter* application is demonstrating many advantages of the IoT approach, esp. increase of process efficiency due to an automated reading of measured consumption data and associated billing without any human interaction. But at the same time, it also explains why IoT security measures are required to protect customer privacy and ensure bullet-proof supply of critical consumables like electricity. For the energy provider, the smart meters minimize risk of electricity theft. For the energy consumer, the smart meter IoT devices and associated services increase awareness and adjust consumption habits. For

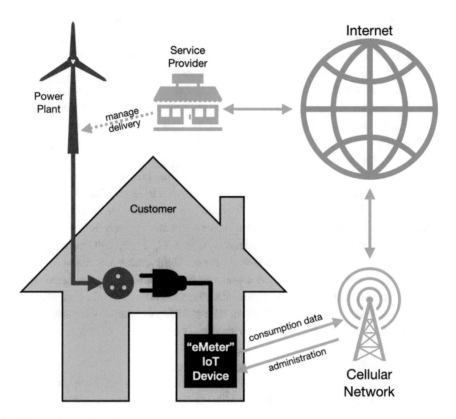

Fig. 2.12 "eMeter" project

example, an electric car can be re-charged at a cheap night rate. But household consumption data also contains private information requiring protection against eavesdropping and misuse. In many countries, this topic is handled by national laws aiming at protection of citizen data privacy. On top of that, smart meter deployment (i.e., refurbishment of existing meter infrastructure) is handled as a mission-critical national project which is supervised in their respective national administrations.

- **Motivation, attack scenarios, potential damage:**

  o Reliable provision of energy is a fundamental governmental objective and a potential target for politically motivated attacks, by terrorists or foreign nations. Denial-of-Service (DoS) attacks against the smart grid infrastructure can be used for this purpose. Compared with other smart grid elements, attacking each single smart meter device offers less efficiency, but must be considered as a thread option.

o As a reference for estimation of potential damage, the smart meter project in Germany can be used: in Germany, the national energy consumption (128 terawatt hours in 2018 [17]) indicates the size of damage in case of a total shutdown of the smart grid infrastructure. With 41 million **private households** and a 1kWh consumer price of around 30 cents, we end up with a total business value of 38 billion EUR in year 2018. Looking at single smart meter installations with an average consumption of 3100 kWh p.a. each household is facing an **annual bill of roughly 1000 EUR**. This huge market is attracting various attackers from different perspectives and with different objectives.

o But for this risk assessment exercise we focus on private households which are motivated to **reduce the invoiced amount of their electricity bill**. For this attacker profile, **physical tampering** attempts against an installed *eMeter* device is feasible. This means that all options described in Sect. 2.4 are available to the attacker.

o Expected skills of typical local attackers (see Sect. 2.6.2) will probably be rated level-1, but level-3 or even level-4 attackers might evaluate vulnerabilities in the background and prepare instructions how to level-1 attackers can execute **sophisticated local attacks**—and share them with interested consumers. In particular, design has to resist attempts to modify sensor data or to inject fake sensor data. Prevention of cloning resp. relocation of *eMeter* devices is crucial.

- **Risk scorecard:**
- IoT security concept (countermeasures): Protection level-3 will be required to handle elaborations by expert hackers who are offering instructions for sophisticated physical attacks to consumers.

o In order to **complicate pentesting** (refer to Sect. 2.6.2), an *eMeter* device should not be available as a commercial product, i.e., attackers cannot easily get hold of a sample. Instead, *eMeters* are ceded by the service provider to each of their customers during the agreed contract period (i.e., device rental instead of sales). Refer to Sect. 4.1.

o **Local physical attacks are a major thread** against *eMeters*, but mechanical protection of an *eMeter* case does not stop attackers from intrusion and tampering internal hardware. But **remote integrity checking** by the operator can detect device tampering, e.g., manipulation of components or modification of embedded IoT application software. See Sect. 4.2.3 for further information. If verification of device integrity fails, the operator will not accept IoT data from this *eMeter* device anymore and take action to replace it, for example.

o An efficient countermeasure against detection of mechanical intrusion are **sensors for light, temperature, humidity** in combination with an **event logging function**. Reported parameters can indicate opening of the eMeter housing (see Sect. 4.1.3).

Table 2.7 Risk scorecard for "eMeter"

STRIDE Thread (see Sect. 2.3)	Relevant ?	Attack Scenario (see text)	Potential damage	protection level (see Sects. 2.6.2 and 2.7)	Security items (see Sect. 2.5)	Countermeasures (see text)
S	yes	Identity theft, device cloning	Supply fee	**3**	2, 5	Strong authentication and identity management
T	yes	Physical software and data tampering	Supply fee		3, 4, 5, 8	Integrated data processing and storage, remote integrity checking
R	yes	Denial of consumption	Billed supply fee		2, 5	Non-Repudiation via signature of registered user
I	yes	Eavesdropping of payload IoT data	Violation of national privacy laws (GDPR)		5	Data encryption
D	yes	Device IP address penetration	Instability of national supply		3	Private device IP address
E	no	n/a	n/a			

o As a starting point for remote device integrity checks, only authorized staff should be allowed to **install and freeze configuration of each *eMeter* device**. Final step of the installation procedure will collect relevant system information, create a corresponding hash value and, for example, digitally sign it using a key card of the service technician. Hash and signature will be used as reference for future integrity checks.

o As a level-3 foundation to protect confidentiality and integrity of data transmission, **public-key cryptography and an associated infrastructure (PKI)** for key management has to be used. For further information refer to Sect. 3.3.

o Device identities are based on a certified user registration process and a trusted third party (certification authority "CA") to manage certificates for used keys.

o **Strong device authentication and a message authentication code (MAC)** are used to for a reliable transmission of consumption data to the service provider. See Sect. 3.4 for further information.

o For a solid **secure communication channel** a standard transmission protocol supporting symmetric cryptography like TLS should be used. Temporary keys are derived from device identities for each session. See Sect. 3.6 for in-depth information.

o A key countermeasure against level-3 identity spoofing and code protection is an **integrated key storage and processing approach**, i.e., inside the same chip. Suitable hardware solutions are explained later in Sect. 4.3.

o As a side-effect from strong device authentication, providers are protecting themselves against repudiation of consumers, if a device is digitally **signing consumption data on behalf of its user**. This is a **non-repudiation approach**, i.e., a consumer cannot successfully dispute its ownership. Used identity association requires a decent registration procedure and an appropriate legal foundation (e.g., via supply contract).

o Another side-effect of applied cryptographic measures is an encrypted transmission of private data via Internet. This will ensure compliance with **national GDPR obligations** (see Sect. 2.9).

o Device identities as well as IP addresses are managed by a dedicated PKI infrastructure. Device interaction as well as device configuration is maintained exclusively by the selected service provider (see Fig. 2.12). **Denial-of-Service (DoS) attacks** can be avoided by keeping device IP addresses confidential (or even change them occasionally). In case of doubt, the operator can request a "sign-of-life" at any time (see Sect. 4.2.1)

Note: Our description of sample products to illustrate the risk assessment process are closely reflecting the actual IoT market situation, but they do not exist in reality. In particular, this applies to *eMeter* which is describing a generalized and slightly simplified product. For a set of documents reflecting real-world security requirements of a national smart meter deployment project, see [18] for specifications of smart meter products and infrastructures in Germany, as an example.

2.9   National Regulations and Legal Standards

Cyberattacks against IoT ecosystems are threatening individuals and corporations which are also taxpayers and voters. It is not a surprise that governments are promoting utilization of IoT security, some are even enforcing national standards for this purpose. Usually, national authorities try to avoid market regulations because economic creativity might suffer from applied rules. But, for example, on May 12, 2021, US President Joe Biden signed an executive order titled **"Executive Order on Improving the Nation's Cybersecurity"** [19] which is specifically aiming at improved national cybersecurity efforts. Goal is to ensure better protection against increasingly sophisticated malicious cyber campaigns causing serious damage. First of all, this initiative calls for actions to modernize cybersecurity in the federal government itself in order to make governmental IT systems more resistant against external attacks. Biden states: "… the prevention, detection, assessment, and remediation of cyber incidents is a top priority and essential to national and economic security. The Federal Government must lead by example." And Biden is asking for "bold" changes instead of "incremental improvements" targeting the public sector as well as the private sector. For this purpose, the $70 billion IT purchasing power of the US government will be used to move market players accordingly.

First of all, manufacturers of IT equipment as well as ICT service providers and software suppliers selling to federal organizations will be affected. But besides IT, Biden also states that "the scope of protection and security must include systems that process data (information technology (IT)) and those that run the vital machinery that ensures our safety (operational technology (OT))." OT is a term which was originally referring to industrial equipment for control and monitoring purposes using dedicated interfaces and communication protocols. Today, IoT devices are increasingly replacing traditional OT devices, i.e., Mr. Biden is indirectly addressing his call also to the Internet of Things. On top of federal procurement, he is asking the National Institute of Standards and Technology (NIST) to "initiate pilot programs informed by existing consumer product labeling programs to **educate the public on the security capabilities of Internet-of-Things (IoT) devices**". For this purpose, the NIST is supposed to "identify IoT cybersecurity criteria for a consumer labeling program" to be used by manufacturers "to inform consumers about the security of their products". This order clearly indicates that the US government has identified a need to create better citizen awareness of potential risks when using IoT applications and devices. And it follows the "Internet of Things Cybersecurity Improvement Act of 2020" which was published before on Dec. 4, 2020 [20].

For manufacturers of IoT devices, these are must-read documents: **NISTIR 8259— "Foundational Cybersecurity Activities for IoT Device Manufacturers"** and **NISTIR 8259A—"IoT Device Cybersecurity Capability Core Baseline"** [21]. They are providing general guidance on how to secure IoT devices. In particular, NISTIR 8259A describes a set of cybersecurity capabilities, as a default for minimally securable IoT devices. NISTIR 8258A includes several recommendations which have been mentioned in this book (see Sect. 2.5), e.g., to establish a reliable **device identification infrastructure** or to provide a **secure field software update** function. While implementation is not required by law, this guidance likely will be referenced when determining the reasonableness of IoT device security. Device manufacturers, particularly those that sell or seek to sell to the US government, should assume security requirements will become the standard and should take these two guidance documents into consideration when designing and implementing new IoT devices.

In Europe, a joint Cybersecurity Act is establishing an EU-wide framework for a common cybersecurity certification approach. This act is supposed to boost the cybersecurity of digital products and services (including IoT) in Europe (for more information see [22]). On top of this, on the EU has taken action to improve cybersecurity of wireless devices to be sold on the EU market on Oct 29, 2021 [23]). A corresponding extension of the "Radio Equipment Directive (RED)" is specifying legal requirements for safeguards, which manufacturers will have to consider in the design and production of the concerned products. This decision is primarily aiming at better protection of consumer privacy and risk of monetary fraud. This means that manufacturers will have to provide evidence that **access control and authentication mechanisms comply with associated standards—probably starting from mid 2024**.

For IoT manufacturers, European standard ETSI EN 303 645 "Cyber Security for Consumer Internet of Things" provides a set of baseline provisions applicable to all consumer IoT devices [24]. In addition, test specification ETSI TS 103 701 describes how a conformity assessment is performed in a structured and comprehensive way. This will allow supplier organizations such as manufacturers, vendors or distributers to assess the compliance of their devices against ETSI EN 303 645 in self-assessments or via testing labs.

Independently from threads caused by cyberattacks, all over the world we have national regulations aiming at **protection of data privacy**. For example, the European GDPR ("General Data Protection Regulation") also applies to the Internet of Things [25]. GDPR has stipulated the basic rules for processing personal data at a European level. However, the GDPR will obviously be relevant to IoT applications only in case that they actually process personal data. This applies, for example, to IoT ecosystems processing personal data in combination with corresponding data originated from acoustic, optical, or biometric sensors. GDPR requires extra attention by IoT applications which are used to identify persons and/or detect presence of a person. This data would allow to create sensitive user profiles allowing to determine movement habits or preferences. It is the responsibility of the providers of IoT applications to ensure that GDPR-compliant data protection and security concepts are implemented.

Besides European GPDR [25], in other parts of the world we also have national data privacy regulations are in place, for example:

- CCPA—California Consumer Privacy Act [26]
- LGPD—Brazil's Lei Geral de Proteção de Dados (General Personal Data Protection Act) [27]
- POPIA—South Africa's Protection of Personal Information Act [28].

2.10 Security Evaluation and Certificates

Manufacturers and service providers with no specific IoT security experience can benefit from external expertise in this particular field. Traditionally, companies focus on addressed use cases, application-specific functions, and unique features—rather than IoT security. But with billions of deployed IoT devices ongoing fueling the worldwide digital transformation, data security and privacy considerations will have a strong impact on product requirements (recall Sect. 2.1). This means that IoT security will be a primary ingredient and should be implemented in the early stages of the product development lifecycle. To do this, device makers might need to change their mindset and adapt their development processes accordingly because required skills are not available in-house. So, at least as a starting point, an **independent security design consultancy** can make a difference and will provide extra confidence that a planned IoT product will meet market expectations.

Avoiding **cost of failure** is another good reason for an independent **risk assessment and product certification**. As already mentioned before, according to a survey of insurer Hiscox [17], the median cost of a cyber breach in 2020 was 57,000 USD. Manufacturers need to take a proactive approach to security, so they are not creating unexpected financial damage due to their inaction. On the other hand, a security certificate is a good instrument for insurers to model risks and offer associated warranties to IoT companies.

On top of that, a successful security evaluation and an associated certificate also works as a selling point—demonstrating company commitment to security. An IoT security certificate provides evidence to concerned customers about the **trustability and quality of an IoT product**.

In general, the term "security evaluation" refers to a comprehensive expert review of a security concept and implementation. In addition, guidance for security hardening is offered. This service is offered by many laboratories all over the world (seeTable 2.8).

As a common practice today, a security evaluation of an IoT device is used determine its degree of compliance with a given security standard or specification. In case of a successful evaluation by a well-recognized test house, compliance will be certified, i.e., a corresponding **security certificate** will be issued. In future, this requirement will apply to most commercial IoT devices (see Sect. 2.9). But for the time being, product certification usually is mandatory for **governmental markets** and for mission-critical deployments under supervision of national authorities, e.g., a countrywide roll-out of smart meters to consumer households. In fact, a CC evaluation is common practice for many **national smartcard projects** used for identification purposes, e.g., for national rollouts of ID cards, electronic passports, but also for health cards or tachograph cards for use in supervised infrastructures. But some IoT deployments deserve special attention by national authorities already today. For example, public energy supply infrastructures create a multi-billion citizen cash flow and require. The nation-wide smart meter roll-out for new-generation German smart grid infrastructure is asking for a **Common Criteria security certificate** for the so-called "Smart Meter Gateway (SMGW)" to be deployed in households and industrial sites. An embedded smartcard is used as a security module for crypto operations and secure storage, this smartcard module will have to be CC-certified at a very high level in order to resist level-4 attacks (see Sect. 2.7).

2.10.1 Schemes

The **Common Criteria (ISO/ISO 15408)** evaluation scheme (short: "CC") was developed in the mid-1990s by Canada, France, Germany, the UK, the USA, and the Netherlands to creating a standard way to specify a computer product security objective and a standard way for security labs to evaluate the products and determine if they actually meet these security objectives. Traditionally CC has been the main international certification program for IT products. For the time being, CC is also the most relevant scheme for IoT products. But in the meantime, IoT project owners can choose from several cybersecurity standards and industry-driven certification schemes. Of course, all of them are used as evidence for good quality IoT security and will be valuable certificates, but some new evaluation methodologies are less stringent and require less efforts by the manufacturer—compared to the Common Criteria approach. Refer to extra Sect. 2.10.2 for detailed information about CC. A newer scheme called **SESIP** (Security Evaluation Standard for IoT Platforms) is closely following CC principles but focuses on security requirement of IoT devices. Details are explained in an extra Sect. 2.10.3 below.

Alternate security evaluation schemes are.

- **ISO/IEC 27,001** for information security management
- **IEC 62,443** for Industrial Automation and Control Systems (IACS), is the primary international reference framework for cyber security in industrial systems, specifying a series of anti-cyber-attack measures
- **ETSI EN 303 645** for consumer Internet of things (IoT) devices
- **PSA Certified** label is aiming at IoT hardware and software based on Platform Security Architecture (PSA) specifications by ARM (see [29]). This certification approach is verifying implementation of several security goals (incl. device identity, secure storage, boot, lifecycle, etc.) and a "Root-of-Trust" foundation ("PSA RoT") for confidentiality and integrity integrated into a piece of silicon. of The PSA framework includes different levels of trust, with each level containing a different level of assessment, with progressively increasing security assurances.

  o PSA Certified Level 1 is aiming at chip vendors, software platforms and device manufacturers. The certification consists of questions, document review and an interview by one of the certification labs, i.e., this certification level is based on **self-assessment**
  o PSA Certified Level 2 is a fixed time, test laboratory based, evaluation of the PSA-RoT. It is aimed at IoT devices that need to protect against scalable **software attack**. It is verifying proper isolation of a secure versus a non-secure processing environment, a secure boot process incl. integrity check, secure firmware update, secure storage (of assets, keys), secure debug ports, proven cryptography.
  o PSA Certified Level 3 adds protection against **physical attacks** as well as side-channel attacks and operation outside specified operating conditions. A hands-on security evaluation of a sample product will be performed by a lab.

Fig. 2.13 Security
certifications

- **ioXt Alliance** certificates are used for a large variety of IoT products from low-level components (e.g., a network module), mobile apps (e.g., for home automation) to connected IoT equipment like air conditioners—with guidelines for the appropriate level of security needed for a specific product. For IoT hardware products (devices, chips), the "ioXt 2021 Base Profile Scheme" specifies a couple of basic security requirements like securing external interfaces, provide proven cryptography, offer secure software update, run only verified software, a security expiration date. allows both self-certification as well as lab certification (for more info see [30]).
- **GSMA IoT Security Assessment** for secure design of cellular IoT services and a mechanism to evaluate security measures (for more info see [31]).
- **FIPS 140** and ISO/IEC 19,790 for cryptography modules

Table 2.8 provides a (non-exhaustive) list of companies providing security evaluation and/or certification services all over the world—covering mentioned cybersecurity schemes.

Common Criteria certification is a well-known and globally accepted approach for IT products. For IoT products CC also becomes mandatory in more and more countries due to governmental regulations, esp. for mission-critical public deployments. In this case, and as a prerequisite to qualify as a potential supplier, manufacturers might have to follow dedicated CC rules (i.e., a protection profile) anyway. But independent from access to these markets, a security certificate demonstrates expertise and commitment to IoT security. In fact, certified IoT products will gain extra market recognition and customer trust. In case a CC certificate is not mandatory, manufacturers can also consider alternate to apply an alternate evaluation scheme how to certify their IoT products, e.g., a SESIP evaluation [32], at level 1 (or PSA Certified [28], if applicable) or GSMA IoT [31], ioXt [29]. and ETSI EN 303 645 [25].

Table 2.8 Security evaluation labs

Name	Locations	Schemes	Website URL
7layers	Germany, France, Beijing	ioXt, GSMA IoT, IEC 62,443	https://www.7layers.com
Applus	Spain, France, UK, Germany, Mexico, Canada, Chile Colombia, China, Norway, USA, etc	CC, GSMA IoT, PSA Certified, SESIP	https://www.appluslaboratories.com
atsec	Texas/USA, China, Italy, Sweden, Germany	CC, FIPS-140	https://www.atsec.com
Brightsight	Netherlands, Spain, France, China, Austria	SESIP	https://www.brightsight.com
CAICT	China	PSA Certified	http://www.caict.ac.cn
CCLab	Hungary	CC, IEC 62,443	https://cclab.com
Corsec	USA	CC, FIPS-140	https://www.corsec.com
DEKRA	Germany	CC, IEC 62,443, ETSI EN 303 645, FIPS-140, ioXt, GSMA IoT	
ECSEC	Japan	CC, SESIP (on the way)	http://www.ecsec.jp/english/index.html
ITSC	Japan	CC, FIPS-140, ISO/IEC 19,790	http://www.itsc.or.jp/en/
JTSec	Spain	CC, IEC 62,443	https://www.jtsec.es
Kiwa	Netherlands	ETSI EN 303,645	https://www.kiwa.com/nl/en/themes/cyber-security/
KOSYAS	South Korea	CC	https://www.kosyas.com
onward	Guangzhou-China, Taipei, Tokyo	CC, FIPS-140, GSMA IoT, IEC 62,443, ioXT	https://www.onwardsecurity.com
Red Alert Labs	France	CC, FIPS-140, IEC 62,443, ISO27K, ETSI EN 303,645, IoXT, GSMA IoT	https://www.redalertlabs.com
Riscure	Netherlands, San Francisco, Shanghai	CC, PSA Certified, SESIP	https://www.riscure.com
Secura	Netherlands	CC, ETSI EN 303,645, GSMA IoT	https://www.secura.com

(continued)

Table 2.8   (continued)

Name	Locations	Schemes	Website URL
Serma	France	CC, GSMA IoT, ETSI EN 303,645, IEC 62,443, SESIP (on the way)	https://www.serma-saf ety-security.com
SGS Brightsight	France, Spain, Netherlands, Austria, Beijing, Taipei	CC, PSA Certified, ioXt, GSMA IoT	https://www.brightsight. com
Teron Labs	Australia	CC	https://www.teronlabs. com
TÜV TRUST IT	Austria	ISO/IEC 27,001	https://it-tuv.com/en

In particular, this applies to manufacturers of IoT components with an uncertain CC roadmap. In fact, due to the formal and stringent nature of Common Criteria, the path to CC certification is painful esp. for CC novices and a significant investment (time, cost). SESIP scheme is quite new but will probably evolve as an accepted and efficient CC alternative for IoT products.

But on the other hand, security components (chips, embedded software) aiming at tamper-proof IoT devices will have to compete against CC-certified products originated from manufacturers with a proven smart card track record. In particular, this applies to secure IoT solutions based on smart card technology. We will take a close look at available options later in Sect. 4.3. But in the end, customer expectations and project requirements will determine which security certificate a manufacturer needs to provide for qualification as a supplier. And, for the time being and, in most cases, this will be a CC certificate.

2.10.2   Common Criteria

Common Criteria is an international standard (ISO/IEC 15,408) for computer security certification. Common Criteria (short: "CC") is a process how to specify functional security and assurance requirements for IT resp. IoT products. In fact, CC certificates are internationally accepted [33] and supported by national governments. The process is initialized by a user or user community requiring a secure product for a certain use case. Typically, this is a governmental institute or organization which can formally specify security requirements as a **Protection Profile (PP)**. Each PP is addressing a specific use case scenario or application, e.g., a machine-readable travel document (ePassport) or a smart meter. A PP is **implementation-independent**, i.e., it does not determine technical means how to meet specified security requirements. A PP may cover hard- and software components and contains

- **Security Problem Definition** incl.
  - Threats,
  - Assumptions,
  - Organizational security policies,
- **Security Objectives**
- **Security Requirements** incl.
  - Functional requirements (SFRs)
  - Assurance requirements (SARs) incl. a target evaluation assurance level (EAL)

All existing PPs are listed on the official CC website [34]. This repository is containing hundreds of security products from all over the world, a Korean single-sign-on, Turkish smart meter, French smartcard reader, German USB storage device, Japanese ePassport, etc. Smartcard products and other secure hard-/software systems like voting machines, vehicle tachographs or smart meters are dominating the list, but you also find software-only objects (e.g., an operating system or a web browser) or entities (e.g., a certification authority CA) or infrastructures.

For manufacturers, usually the Common Criteria certification process begins by contacting the national Certification Body (CB), see worldwide list [35]. The Certification Body is the entity issuing the final certificate when the evaluation is completed, so they are ultimately accountable for the quality of the evaluation. The assessment is not performed directly by the CB, it is necessary to hire the services of an accredited laboratory of his choice (see [36]). Once the lab has performed the evaluation and after identified vulnerabilities have been corrected in cooperation with the manufacturer, the lab will send to the CB an "Evaluation Technical Report" with a "Pass" result, so finally the certificate will be published and may be enforced worldwide.

Test candidates submitted to the lab are called **Target of Evaluation (ToE)**. In order to feed the evaluation work with more detailed information, the manufacturer has to deliver an **implementation-specific** document called **Security Target (ST)**. The ST explains (to the test lab) how the PP requirements have been implemented and what is to be evaluated. Figure 2.14 is illustrating the CC certification process and stakeholders.

A PP is containing a set of SFRs which are used to specify security enforcements by the ToE incl. mandatory functions to be implemented. For example, for a **digital tachograph** for vehicle speed tracking according to EU regulations, an extra **CC-certified sensor unit** must be used. Security requirements are specified in protection profile "Digital Tachograph—Motion Sensor (MS PP)", see [36]. This sensor works in conjunction with other building blocks of the tachograph: a vehicle unit (VU), a tachograph card (TC) and an external GNSS facility (see Fig. 1 of [37]). The motion sensor monitors the vehicle gearbox and provides signals to the vehicle unit that are representative of vehicle movement and speed. First of all, the PP defines the security problem, i.e., assets, attacker profile and threats to be considered by the manufacturer for implementation. Assets to be protected by the ToE are measured motion data, stored data (identification data, keys, and audit records), embedded software as well as supporting hardware. Potential attackers are

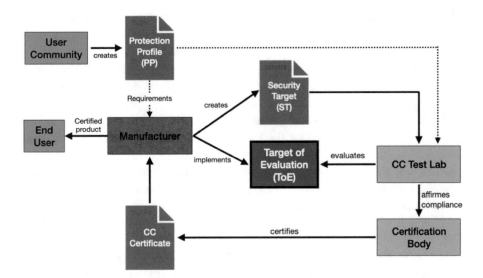

Fig. 2.14 CC certification process overview

described as "a human, or process acting on their behalf, located outside the ToE. For example, a driver could be an attacker if he attempts to interfere with the motion sensor. An attacker is a threat agent (a person with the aim of manipulating user data, or a process acting on their behalf) trying to undermine the security policy defined by the current PP, especially to change properties of the maintained assets. The attacker is assumed to possess at most a **high attack potential**".

Threats to be averted by ToE are specified as.

- **Access control**—A vehicle unit or other device (under control of an attacker) could try to use functions not allowed to them, and thereby compromise the integrity or authenticity of motion data.
- **Design knowledge**—An attacker could try to gain illicit knowledge of the motion sensor design, either from manufacturer's material (e.g., through theft or bribery) or from reverse engineering, and thereby more easily mount an attack to compromise the integrity or authenticity of motion data.
- **Environmental attacks**—An attacker could compromise the integrity or authenticity of motion data through physical attacks on the motion sensor (thermal, electromagnetic, optical, chemical, mechanical).
- **Modification of hardware**—An attacker could modify the motion sensor hardware, and thereby compromise the integrity or authenticity of motion data.
- **Interference with mechanical interface**—An attacker could manipulate the motion sensor input, for example, by disconnecting the sensor from the gearbox, such that motion data does not accurately reflect the vehicle's motion.

- **Interference with motion data**—An attacker could add to, modify, delete, or replay the vehicle's motion data, and thereby compromise the integrity or authenticity of motion data.
- **Access to security data**—An attacker could gain illicit knowledge of secret cryptographic keys during security data generation or transport or storage in the equipment, thereby allowing another device to be connected.
- **Attack on software**—An attacker could modify motion sensor software during operation, and thereby compromise the integrity, availability, or authenticity of motion data.
- **Invalid test modes**—The use by an attacker of non- invalidated test modes or of existing back doors could permit manipulation of motion data.
- **Interference with power supply**—An attacker could vary the power supply to the motion sensor, and thereby compromise the integrity or availability of motion data.

In order to prevent attackers to put these threads into action, a set of SFRs must be implemented and verified. Some of them are extracted from the sensor PP and listed in Table 2.9 for illustration purposes.

On top of SFRs, the PP will specify assurance requirements (SARs) the ToE will have to satisfy in order to meet the security objectives for the ToE. A related numerical rating called "Evaluation Assurance Level" (EAL1 through EAL7) reflects how thoroughly products will be tested by the lab, i.e., a successful evaluation proofs the quality of implementation. Each EAL corresponds to a **package of SARs** which covers the complete development of a product, with a given level of strictness. For example, "EAL4 +" means that—all assurance components within mentioned SAR package—have been verified at level 4 whereas the "+" indicates that for at least one assurance component a higher level has been applied. SARs involve **design documentation, analysis and functional or penetration testing** (see Sect. 2.4). The highest level provides the highest guarantee that the security features are reliably applied. Table 2.10 outlines the meaning of each EAL rating.

Table 2.9 Sample SFR list

SFR	Rationale
FPT_PHP.2	Requires that attempts at physical tampering are detected, and that, if the case is designed to be opened, an audit record is generated
FPT_TST.1	Self-tests help to ensure that the TOE is operating correctly
FTP_ITC.1	Requires use of a secure channel for communication with the VU
FDP_SDI.2	Requires the TOE to monitor stored data for integrity errors
FAU_GEN.1	Audit records are stored when attempted physical tampering is detected
FDP_ACC.1	Require that unauthenticated software is not accepted

Table 2.10 EAL levels

EAL Level	Description
EAL 1	Functionally tested
EAL 2	Structurally tested
EAL 3	Methodically tested and checked
EAL 4	Methodically designed, tested and reviewed
EAL 5	Semi-formally designed and tested
EAL 6	Semi-formally verified design and tested
EAL 7	Formally verified design and tested

Although assurance requirements for each product and system are the same, functional requirements differ. Functional features are created in the ST document, which is specifically tailored for each product's evaluation. A **higher EAL does not indicate a higher level of security than a lower EAL** because they may have different functional features in the STs.

Entry point for all manufacturers of IoT products interested in the Common Criteria standard is www.commoncriteriaportal.org. In this repository, latest versions of the standard as well as current PPs and all certified products (incl. each certificate and ST of the ToE) are listed. A good CC tutorial for IoT project managers presented by secure software company Trustonic (www.trustonic.com) and CC test lab Riscure is called "Security certification considerations when choosing a secure product" and can be found here under [38].

2.10.3 SESIP

SESIP (Security Evaluation Standard for IoT Platforms, see [39]). SESIP has evolved out of Common Criteria and follows the same principles. Like CC, SESIP relies on a strong formalism but focusses on IoT devices. For example, a threat model adapted to the IoT ecosystem. In addition, reusability is an essential objective of SESIP in order to reduce evaluation cost. Another goal of SESIP is to make sure that the results of an evaluation are accessible and exploitable by security-proficient developers without the need to be evaluation specialists. SESIP also offers an option for security self-assessment. SESIP is quite new with first certificates issued in 2019, see [40] for a complete list of valid certificates which have been issued so far.

SESIP Profiles and SESIP Mappings are documents intended to support the usability of the SESIP methodology in the field. The SESIP methodology allows the definition of a security profiles which is generic to an IoT component, e.g., an MCU. These are called SESIP Profiles and are equivalent to CC Protection Profiles. A SESIP Profile document is a **generic SESIP Security Target** defining the SESIP requirements in terms of security features and evaluation activities. SESIP profiles are supposed to make sure that any SESIP-certified IoT component (e.g., a security MCU or a secure element) implements an expected minimum set of features, reaching a targeted security level. In addition, as the same set of features is assessed based on the same evaluation activities, this allows comparability between SESIP evaluations of different IoT components.

SESIP allows the functionality range typically required by secure IoT products from the underlying platform to be evaluated to one of five hierarchical assurance levels. This way, a platform can be evaluated once, and many secure products can be built on top of it without the need to re-evaluate the platform over and over. For more information about this approach see document "Security Evaluation Standard for IoT Platforms (SESIP)" ([41]).

There are three primary assurance levels in SESIP and two extended levels, which are:

o SESIP Assurance Level 1 (SESIP1) is a **self-assessment**-based level that provides a basic level of assurance. Only minimal verification by the evaluator is required.
o SESIP Assurance Level 2 (SESIP2) is a **black-box penetration testing** level that provides a moderate level of assurance. SESIP2 provides significantly more assurance than SESIP1 by requiring a vulnerability analysis and actual penetration testing on the platform by an evaluator.
o SESIP Assurance Level 3 (SESIP3) is a traditional **white-box vulnerability analysis** that provides a substantial level of assurance. This evaluation is structured around a time-limited source code analysis combined with a time-limited penetration testing effort.
o SESIP Assurance Level 4 (SESIP4) **complements an existing CC** evaluation of the candidate (ToE) which is compliant with a list of SESIP4 assurance components incl. AVA_VAN.4 (Methodical vulnerability analysis and resistance against **moderate attack potential**). SESIP4 is raising the required product resistance level, adding design implementation analysis, and strengthening the production environment assessment
o SESIP Assurance Level 5 (SESIP5) is the same as the standard high assurance level currently being used for smartcards, secure elements, e-passports, etc. It provides a **very robust defense against very advanced threats**. SESIP5 complements an existing CC evaluation of the candidate (ToE) which is compliant with a list of SESIP5 assurance components incl. AVA_VAN.5 (Advanced Methodical vulnerability analysis and resistance against **high attack potential**).

References

1. IoT Analytics. State of IoT 2021: Number of connected IoT devices. Retrieved 28 January, 2022, from https://iot-analytics.com/number-connected-iot-devices/.

2. Treatpost: IoT Attacks Skyrocket, Doubling in 6 Months. Retrieved 28 January, 2022, from https://threatpost.com/iot-attacks-doubling/169224.
3. IoT Analytics. Top 10 IoT applications in 2020. Retrieved 22 January, 2022, from https://iot-analytics.com/top-10-iot-applications-in-2020.
4. The STRIDE Threat Model. Microsoft. Retrieved 23 January, 2022, from https://msdn.micros oft.com/en-us/library/ee823878(v=cs.20).aspx.
5. u-blox. SARA-R5 series AT Commands Manual. Retrieved 28 January, 2022, from https://www.u-blox.com/en/docs/UBX-19047455.
6. Jean-Georges Valle. (2021, April). Practical Hardware Pentesting. Packt Publishing,. ISBN: 9781789614190.
7. Aaron Guzman, Aditya Gupta. IoT Penetration Testing Cookbook. Packt Publishing 2017. ISBN: 9781787280571.
8. Georgia Weidman. Penetration Testing—A Hands-On Introduction to Hacking. No Starch Press 2014. ISBN: 9781593275648.
9. Keng Tiong Ng. Manual PCB-RE. 2021. Amazon Digital Services LLC—KDP Print US, ISBN: 9798716998513.
10. NXP Semiconductor. I2C-bus specification and user manual, Rev. 7.0 — 1 October 2021. Retrieved 15 February, 2022, from https://www.nxp.com/docs/en/user-guide/UM10204.pdf.
11. Payatu. Introduction to Fault Injection Attack (FI). Retrieved 30 March, 2022, from https://pay atu.com/blog/asmita-jha/fault-injection-basics.
12. STMicroelectronics. AN5156 Introduction to STM32 microcontrollers security. Retrieved 10 June, 2022, from https://www.st.com/resource/en/application_note/an5156-introduction-to-stm32-microcontrollers-security-stmicroelectronics.pdf.
13. Payatu. Introduction to Side Channel Attacks (SCA). Retrieved 30 March, 2022, from https://payatu.com/blog/asmita-jha/side-channel-attack-basics.
14. NewAE Technology Inc. ChipWhisperer. Retrieved 30 March, 2022, from https://www.newae.com/chipwhisperer.
15. Tutorialspoint. Advanced Encryption Standard (AES). Retrieved 30 March, 2022, from https://www.tutorialspoint.com/cryptography/advanced_encryption_standard.htm.
16. HISCOX Cyber Readiness Report 2020. Retrieved 28 February, 2022, from https://www.hiscox group.com/sites/group/files/documents/2020-06/Hiscox-Cyber-Readiness-Report-2020.pdf.
17. GMI. Smart Electric Meter Market. 2021. Retrieved 28 September, 2021, from https://www.gmi nsights.com/industry-analysis/smart-electric-meter-market.
18. Statistisches Bundesamt. Energy Consumption. 2022. Retrieved 15 February, 2022, from https://www.destatis.de/EN/Themes/Society-Environment/Environment/Material-Energy-Flows/Tab les/electricity-consumption-households.html.
19. Bundesamt für Sicherheit in der Informationstechnik (BSI). Smart Meter Gateway. Retrieved 22 February, 2022, from https://www.bsi.bund.de/DE/Themen/Unternehmen-und-Organisat ionen/Standards-und-Zertifizierung/Smart-metering/Smart-Meter-Gateway/smart-meter-gat eway_node.html.
20. The White House, Joe Biden. Executive Order on Improving the Nation's Cybersecurity. May 12, 2021. Retrieved 28 September, 2021, from https://www.whitehouse.gov/briefing-room/pre sidential-actions/2021/05/12/executive-order-on-improving-the-nations-cybersecurity/.
21. IoT Cybersecurity Improvement Act of 2020, US Public Law No: 116–207. April 4, 2020. Retrieved 26 September, 2021, from https://www.congress.gov/bill/116th-congress/house-bill/ 1668/text.
22. NIST National Institute of Standards and Technology. NISTIR 8259 Foundational Cybersecurity Activities for IoT Device Manufacturers. Retrieved 28 February, 2022, from https://csrc.nist.gov/publications/detail/nistir/8259/final.

23. European Commision. Shaping Europe's digital future. Retrieved 28 January, 2022, from https://digital-strategy.ec.europa.eu/en/policies/cybersecurity-act.
24. European Commision. Commission strengthens cybersecurity of wireless devices. Retrieved 28 February, 2022, from https://ec.europa.eu/commission/presscorner/detail/en/ip_21_5634.
25. ETSI European Telecommunications Standards Institute. EN 303 645. Cyber Security for Consumer Internet of Things: Baseline Requirements. Retrieved 8 March, 2022, from https://www.etsi.org/deliver/etsi_en/303600_303699/303645/02.01.01_60/en_303645v020101p.pdf.
26. European Union GDPR.EU. Complete guide to GDPR compliance. Retrieved 28 January, 2022, from https://gdpr.eu.
27. State of California Department of Justice. California Consumer Privacy Act (CCPA). Retrieved 28 January, 2022, from https://oag.ca.gov/privacy/ccpa.
28. Presidência da República. Lei Geral de Proteção de Dados Pessoais (LGPD). Retrieved 28 January, 2022, from https://lgpd-brazil.info.
29. POPIA. Protection of Personal Information Act (POPI Act). Retrieved 28 January, 2022, from https://popia.co.za.
30. PSA Certified. IoT Security Framework and Certification. Retrieved 9 March, 2022, from https://www.psacertified.org.
31. ioXt Alliance. Internet of secure things. Retrieved 9 March, 2022, from https://www.ioxtalliance.org.
32. GSMA IoT Security Guidelines and IoT Security Assessment. Retrieved 9 March, 2022, from https://www.gsma.com/iot/iot-security-assessment.
33. GlobalPlatform SESIP. Retrieved 20 March, 2022, from https://globalplatform.org/sesip.
34. Common Criteria. Arrangement on the Recognition of Common Criteria Certificates. July2, 2014. Retrieved 26 March, 2022, from http://www.commoncriteriaportal.org/files/CCRA%20-%20July%202,%202014%20-%20Ratified%20September%208%202014.pdf.
35. Common Criteria. Protection Profiles. Retrieved 8 March, 2022, from https://www.commoncriteriaportal.org/pps/.
36. Common Criteria. Licensed Laboratories. Retrieved 8 March, 2022, from https://www.commoncriteriaportal.org/labs/.
37. Common Criteria. Certificate Authorizing Schemes. Retrieved 14 March, 2022, from https://www.commoncriteriaportal.org/ccra/schemes/.
38. Federal Office for Information Security (BSI). Digital Tachograph—Motion Sensor (MS PP). Retrieved 12 March, 2022, from https://www.commoncriteriaportal.org/files/ppfiles/pp0093b_pdf.pdf.
39. Trustonic, Riscure. Security certification considerations when choosing a secure product. Retrieved 20 March, 2022, from https://www.trustonic.com/wp-content/uploads/2020/09/Security-certification-considerations-when-choosing-a-secure-product-v2.pdf.
40. GlobalPlatform SESIP. Security Evaluation Standard for IoT Platforms (SESIP) v1.1 | GP_FST_070, June 2021. Retrieved 20 March, 2022, from https://globalplatform.org/specs-library/security-evaluation-standard-for-iot-platforms-sesip-v1-0-gp_fst_070/#.
41. SESIP Certificates. Retrieved 20 March, 2022, from https://trustcb.com/iot/sesip/sesip-certificates/.

Cryptographic Toolkit

3

In a typical IoT system, we have a single IoT server working as a supervisor for multiple IoT devices. Each IoT device is securely connected through an insecure public network using the concept of **end-to-end security**, see Fig. 3.1. This means that it—from a security point of view—it does not matter which technology is used for data transport because protection is covering everything in between two endpoints, i.e., the IoT device and the server. A secure communication channel is protecting against identity spoofing, eavesdropping, or altering data. Communication partners are mutually authenticated, and transmitted data is encrypted. This work is performed exclusively by cryptographic modules which are integrated into both endpoints.

Fig. 3.1 IoT end-to-end security

© The Author(s), under exclusive license to Springer Nature Switzerland AG 2022
K. Heins, *Trusted Cellular IoT Devices*, Synthesis Lectures on Engineering, Science, and Technology, https://doi.org/10.1007/978-3-031-19663-8_3

As already explained in Sect. 1.5, our design focus is on the IoT device because this endpoint typically is the weakest element within the IoT ecosystem. In fact, the IoT device will have to.

- manage mutual authentication with the IoT server,
- encrypt IoT payload data for transmission,
- decrypt incoming control commands or firmware updates,
- store/process/manage corresponding cryptographic keys

The low-level foundation for secure IoT applications is quite the same as for IT equipment. IoT security consolidates methods and best practices from well-established and high-volume IT markets requiring strong protection against misuse and tampering, for use cases like payment, access control, surveillance, asset tracking. In particular, this applies to cryptographic fundamentals, key management, and identity deployment infrastructures which are key technologies for secure IoT devices—requiring reliable mutual authentication of communication partners and bullet-proof data protection.

Encryption is a mathematical method for data (messages or files) to be made unreadable, ensuring that only an authorized person can access that data. Encryption ensures that information stays private and confidential, whether it's being stored or in transit. Encryption uses complex algorithms to scramble data and decrypts the same data using an encryption resp. decryption key which has been exchanged by communication partners before. Only the correct key can convert encrypted text (ciphertext) to plaintext. Any unauthorized access, i.e., by somebody who does not have the right decryption key, will only see a chaotic array of bytes.

Practically, cryptographic methods are addressing main IoT security goals and solve typical communication problems such as to

1. protect message **privacy**, i.e., nobody else is able read contents of message,
2. verify **authenticity** of the sender of a message and its **non-repudiation**, i.e., the sender cannot deny authorship of a sent message,
3. ensure **integrity** of message, i.e., make sure that nobody is able to modify message contents on its way to the recipient.

Problem 1 can be solved by **data encryption**, i.e., sender uses the public key of the recipient → decryption of the received message can be done with the recipient's private key only. Problems 2 and 3 can be solved by use of a **digital signature** or a **MAC (message authentication code)**.

3.1   Symmetric Versus Asymmetric Ciphers

Historically, cryptography started with symmetric algorithms like the substitution cipher, which is based on a shared rule, e.g.

$$(a \rightarrow K, \ b \rightarrow Z, \ c \rightarrow M, \ d \rightarrow G, \ \ldots)$$

Using this rule, a transmitted ciphertext of "GKKZ" would convert into a plaintext of "daab" by the recipient. Today, symmetric ciphers are using a key instead of a substitution rule, but the encryption key still has to be shared because it is used by the sender as well as by the recipient, i.e., the same secret key is used for encryption as well as for decryption. A symmetric algorithm might be superfast and unbreakable, but still suffers from a natural **key management problem**: in order to prevent eavesdropping by anybody else, each secure communication channel between two parties requires a unique key. This means that with n participants you will need to agree and distribute (n-1) keys, i.e., potentially a large number of keys. But even worse, you will have to make sure that each key is delivered to the communication partner in a perfectly safe way. On top of this, a related key management infrastructure should replace keys regularly or in case the key has been compromised.

In order to overcome these drawbacks, Martin Hellman, Ralph Merkle, and Whitfield Diffie presented the idea of public-key cryptography at Stanford University in 1976. Other than symmetric cryptography, public-key (aka asymmetric) cryptography uses a pair of keys which are different but mathematically linked to each other: i.e., a public key, which may be disseminated widely, and a private key, which are known only to the "owner". In a one-to-one electronic communication one of the keys (either public or private—depending on use case) is used by the sender, the other one is used by the recipient. This fundamental difference is illustrated in Fig. 3.2.

Fig. 3.2 Keys for symmetric versus asymmetric encryption

Public-key cryptography has a big advantage when compared to classic symmetric crypto schemes where same unique secret key is used by sender as well as by recipient: effective security requires keeping only the private key as a secret; the public key can be openly distributed without compromising security. As a result, the receiver of a confidential message will have to **deploy his public key** to be used for exclusive data encryption by the sender. This means that key distribution and key storage are much easier to handle as with symmetric cyphers.

Today, the following encryption ciphers are popular for use with commercial IT or IoT applications:

- **Symmetric**
  - **AES (Advanced Encryption Standard)** has been selected by the U.S. government to protect classified information, replacing its predecessor DES which was officially withdrawn in 2005. AES allows you to choose a 128-bit, 192-bit or 256-bit key, making it exponentially stronger than the 56-bit key of DES.
- **Asymmetric**
  - **RSA (Rivest–Shamir–Adleman)** has been publicly described back in 1977. The security of RSA relies on the practical difficulty of factoring the product of two large prime numbers, the "factoring problem". There are no published methods to defeat the system if a large enough key is used, so RSA is still widely used in commercial IT systems.
  - **Elliptic-curve cryptography (ECC)** is a newer approach based on the algebraic structure of elliptic curves over finite fields. ECC allows smaller keys compared to non-EC cryptography.

Though private and public keys are related mathematically, it should not be feasible to calculate the private key from the public key. Public-key cryptography is based on the intractability of certain mathematical problems, i.e., a problem that can be solved in theory, but for which in practice any solution takes too many resources (computing power, time) to be useful [1]. In fact, strength of any public-key algorithm is in designing a relationship between two keys. But even if a cypher is unbreakable by exploiting structural weaknesses in its algorithm, it is always possible to run through the entire space of keys in what is known as a "brute-force attack", i.e., a smart way of guessing the correct key. Since longer keys require exponentially more work to brute force search, a sufficiently long key makes this line of attack impractical. But due to the fact that one key is public, asymmetric crypto algorithms are **more complex and require longer key lengths** compared to symmetric cryptography—at the same level of security. For example, security provided with a 1024-bit key using asymmetric RSA is considered approximately equal to an 80-bit key in a symmetric algorithm like AES [2].

Table 3.1 Key length versus security level

Algorithm	Security level (bit)			
	80	128	192	256
RSA	1024 bit	3072 bit	7680 bit	15,360 bit
ECC	160 bit	256 bit	384 bit	512 bit
AES	80 bit	128 bit	192 bit	256 bit

In order to compare different algorithms, sometimes the so-called **security level** is used. A security level of n bit means that the best-known attack required 2^n steps [2]. For symmetric algorithms this is a natural definition because for them a security level of n is equivalent to a key length of n bit. For asymmetric algorithms, this relationship is not straightforward. Table 3.1 lists recommended key lengths for RES, ECC and AES algorithms at different security levels [2]. It shows that RSA requires very long keys whereas ECC is equivalently secure with shorter keys, requiring only twice the bits as the equivalent symmetric algorithm, i.e., a 256-bit ECC key offers approximately same safety factor than 128-bit AES.

By nature, public-key algorithms offer excellent key management support, so they are suitable for node identification purposes in large and scalable IoT installations. They also provide non-repudiation and message integrity when used with digital signature, so they seem to be a perfect choice for all kind of security problems, but there is a major drawback: encryption of data requires a lot of computing power. A block ciphers like AES are **100...1000 times faster than public-key algorithms**. On top of this, key used for symmetric cryptography typically are shorter and reduce required computational effort which is particularly **beneficial for low-cost and battery powered IoT devices**.

As a consequence, and in order to use best of both worlds, hybrid implementations are used for secure **data transmission protocols**, e.g., for the SSL/TLS protocol (see Sect. 3.6.1). So, for example, a TLS session might use RSA to authenticate partners and then agree on an AES key, which is then used for bulk encryption–decryption of messages within a session.

Mentioned algorithms are used for encryption, but other for our IoT cryptographic toolkit a couple of additional ingredients are required:

- **Hash**. A hash function creates a small message digest which is representing the original message. Hash functions are required for digital signature schemes and for message authentication codes.

- **Key Establishment**. In order to set up a secure channel allowing two parties to communicate confidentially, both parties first have to agree on a shared secret key for encryption/decryption. A famous example is the Diffie–Hellman key exchange. This protocol enables users to securely exchange secret keys even if an opponent is monitoring that communication channel. It is explained in Sect. 3.5
- **Key Generation and Identity Management**. Use of public keys also means that some assurance of the **authenticity of a public key** is needed in this scheme to avoid spoofing. Generally, this type of cryptosystem involves a trusted third party ("certification authority") guaranteeing that a particular public key belongs to a specific person. This topic is handled in next Sect. 3.3.

3.2   Digital Signatures

A digital signature is a key cryptographic tool for IT deployments. Digital signatures are widely used whenever trusted online communication is required, e.g., for eCommerce or legally relevant transactions. A digital signature is used to verify the authenticity of message or any kind of digital data, e.g., document or a software package. It is kind of an electronic version of a handwritten signature which is providing strong evidence who created a data package (user authenticity) and if it has been altered in the meantime (data integrity). In fact, this works with public-key cryptography only, because symmetric keys are shared and—by nature—known by at least two users because the same key is used for encryption as well as for decryption. This means that both parties can use it either way. With public-key cryptography, one part of the key pair is private and unique and will not be shared with anybody. This private key is used to sign a message, and only the key owner knows it and can use it. Consequently, a specific key used for signing a document proves authorship, the key owner cannot deny. This feature is called **non-repudiation** and another characteristic highlight of public-key schemes.

So, the sender's private key is used to sign a message, and all recipients can verify the sender's authenticity with the sender's public key. In order to limit required computational effort to create a digital signature, only the **hash** value of the message is encrypted, not the complete message itself. What does "hash" mean? A hashing algorithm is a mathematical function that condenses data to a fixed size, a hash is a fingerprint of the original data. On top of that a **secure hash** is irreversible and unique. Irreversible means "one-way", i.e., from the hash itself you couldn't figure out what the original piece of data was, therefore allowing the original data to remain secure and unknown. Unique means that two different pieces of data can never result in the same hash value. Today, for digital signatures SHA-2 algorithms are common, e.g., SHA-256 with a hash length of 256 bit.

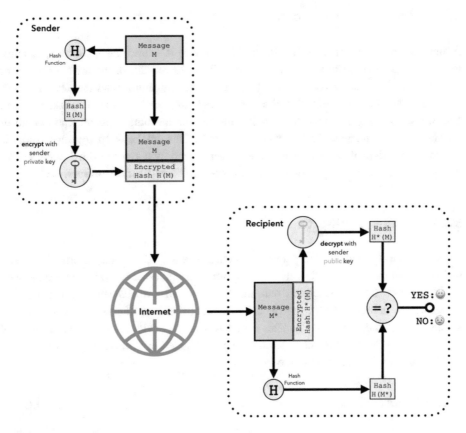

Fig. 3.3 Digital signature—create and verify process

So, **on sender side** (see Fig. 3.3) the message will be hashed, encrypted with sender's private key (aka "signed") and attached to the message before being transmitted via a public (i.e., unsecured) network. In order to verify sender authenticity and message integrity this signature will be decrypted with the sender's public key **on receiver side** [2]. This operation creates H*(M). For verification purposes the received message M* will be hashed using the same hash algorithm as on sender side. This auxiliary hash H(M*) must be identical to the received decrypted hash H*(M). If not equal, very obviously something is wrong, either because

1. used public key on sender side does not match → identity of sender is questionable

or

2. received message is not identical with original message → message content is questionable.

3.3   Identity Management

Assignment of a **unique identifier to each IoT device** is a fundamental security requirement for every IoT installation (refer to Sect. 2.5). As explained, the private portion of
an individual key pair is used for high-quality digital signatures, and the key pair itself
represents the **electronic identity** of a person resp. an associated IoT networked device.
But this works only if a signature verification returns meaningful identification data of
the signee, e.g., a name and relevant contact details. This link between key and identity is
provided by a **digital certificate** which has been issued by a **trusted third party (TTP)**
which has been selected for a deployment, a so-called **certification authority (CA)**.

3.3.1   Digital Certificate

Digital certificates include the public key being certified, identifying information about
the entity that owns the public key, an expiration date and other relevant metadata related
an IoT project. Collection of this data must be done in a trustworthy manner, e.g., by
a registration authority (RA), see Sect. 3.3.2. Finally, the certificate will be created by
signing collected data by the CA. This signature works as evidence during verification of
a signed message or document which has been exchanged by authorized members within
an IT resp. an IoT deployment. See Fig. 3.4.

Fig. 3.4 Digital certificate

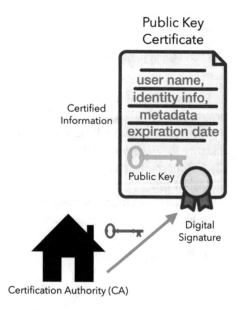

3.3.2 Public-Key Infrastructure (PKI)

For typical IoT applications, esp. for large-scale deployments, management of IoT device identities is a challenge because deployment is an ongoing process throughout the whole lifetime of the application. In general, a suitable infrastructure must prepare for a growing number of users, i.e., new IoT devices are added. Each device identity corresponds to a unique key pair requiring administrational effort while they are operational. person or an object. This why a **public key infrastructure (PKI)** is required for every IoT deployment using public-key cryptography. A PKI is managing certificates and responsible to.

- enter new devices and users according to specific IoT application and deployment needs,
- collect associated user identity information for each new device,
- generate a unique key pair (internal or external) for each new device,
- issue and distribute digital certificates,
- revoke expired certificates or certificates of removed devices,
- deploy updates resp. provide access to an updated list of valid certificates.

Besides offering these fundamental functions, each PKI is based on project-specific policies, roles, and procedures, e.g., how new devices are registered, which metadata is included in certificates, how certificates are distributed, etc. But first of all, a PKI is a dedicated software application, but also offered as a service (see Chap. 4 containing an extra section called "IoT Clouds"). Advanced security requirements might apply to key generation, PKI organization and procedures, esp. in case of governmental projects (see "Security Evaluation and Certificates"), but in most cases the IoT project owner will have several choices how to set up a PKI and related in- and outputs. For example, in addition to a CA for IoT device identities, it might be appropriate to create a 2nd-level CA for users or services, see Fig. 3.5. But it any case, an up-to-date certificate database must ensure that all valid public keys are available to all participants of the IoT deployment, e.g., via central online repository.

Generation, storage, and distribution of key pairs is handled separately requires extra care because of the secret private part. Protection level and infrastructure for key deployment will follow specific needs of the IoT project. Traditionally, a so-called HSM ("hardware security module") is used for this purpose, a tamper-resistant computing device with dedicated cryptographic hardware and protected memory. But in the meantime, also other approaches are available how to generate keys and forward them securely to the CA. For example, a key pair can be generated inside an IoT device with the public part transferred to the CA for certification—without moving the private key, i.e., keeping the secret key inside the device forever. Suitable components will be introduced later in Sect. 4.3.

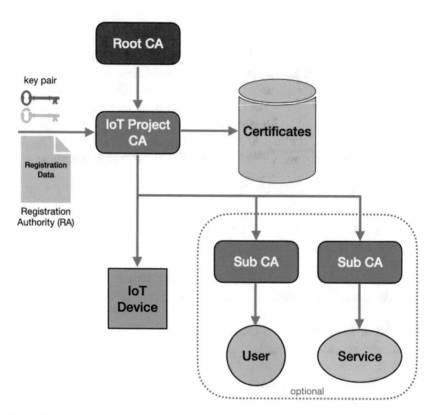

Fig. 3.5 PKI structure

A **registration authority (RA)** is a trusted entity that handles the process of applying for a certificate. In fact, the RA is a PKI role, and some deployments combine RA and CA functions, but in many cases, it makes sense for a PKI to delegate RA role and keep it separately in order to increase flexibility of adding or replacing an RA, if needed. Basically, the RA is responsible for accepting user requests for digital certificates, authenticate applicants and collect pre-defined metadata to be included in the certificate.

For large-scale rollouts like a national citizen ID card, a suitable PKI will be very complex, and RAs will be located all over the country in order to allow citizens to apply locally. But for many IoT deployments, user registration is easier to handle, maybe because affected users are already known, e.g., employees or patients or electricity consumers. In general, PKI implementation and RA structure as well as related processes will depend on the particular IoT use case, on deployment characteristics and how users are involved. Keys and certificates are used by IoT devices "on behalf" of an associated user, not be the user itself. This implies that IoT usage scenarios are different from use of a personal token like a smart card.

For many commercial IoT applications and corresponding deployment, IoT devices will not be "personalized" at all, but have a unique identity based on other data which has been loaded during production. As long as an IoT installation does not have to comply with a given specification, a PKI will have to meet business requirements only and might be tailored to application requirements; and should be as simple as possible. But for sure, each IoT installation will have to consider, specify, and implement all relevant aspects how to manage device identities and private key as well as how to deploy public keys and associated certificates.

For a simple CA and tailored PKI, OpenSSL can be used [3]. It is a toolkit, developed in C, that is included in all major Linux distributions, and can be used both to build your own (simple) CA and to PKI-enable applications (see PKI tutorial at [4]). As a standard format for interchangeable certificates, X.509 [5] is widely used, also for the SSL/TLS secure transmission protocol (see Sect. 3.6.1).

3.4 Message Authentication Codes (MACs)

A Message Authentication Code (MAC) is a cryptographic checksum which is quite similar to a digital signature since a MAC is also providing evidence of message authenticity and integrity. However, unlike digital signatures, MACs are symmetric-key schemes, so they **do not provide non-repudiation**. As a consequence, MACs cannot substitute PK-based digital signatures, but the big advantage of MACs is that they are **much faster than digital signatures** since they are based on either block ciphers or hash functions [2]. Consequently, MACs and digital signatures complement each other in hybrid security solutions for IT and IoT deployments.

Similar to digital signatures, MACs append an authentication tag to a message. MACs can are constructed in essentially two different ways, using a hash function like SHA-x or a block cipher like AES as building blocks. Hash-based MACs are called **HMAC** and became popular because it is used for Transport Layer Protocol TLS. The crucial difference between MACs and digital signatures is that MACs use a shared symmetric key for both generating the authentication tag and verifying it. Rest of the process is equivalent, see Sect. 3.2.

Figure 3.6 is illustrating the HMAC process on sender and receiver side. In short, a message will be hashed **H(M)** and transmit the encrypted **H(M)**. The receiver will create two different hash values:

1. decrypt the received hash value **H*(M)** and
2. repeat the sender hash process using the received message **H(M*)**.

If these two values are different, something went wrong: either wrong key or altered message.

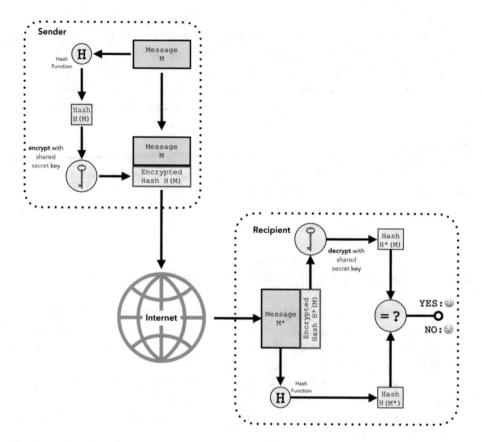

Fig. 3.6 Message Authentication Code (HMAC) process

3.5   Key Establishment

Fundamental IoT security requirements are addressed by symmetric and public-key cryptographic functions:

- Confidentiality by data encryption
- User/Device Authentication by digital signatures (with PKI support)
- Message Authentication by MACs or digital signatures
- Message Integrity by MACs or digital signatures
- Non-repudiation with digital signatures

However, all these cryptographic mechanisms are based on the assumption that keys are properly distributed between the parties involved. Practically, the task of key establishment is one of the most important and often also most difficult parts of a security system [2]. Objective of key establishment is to **share a secret key** between two or more parties. There are two ways how to achieve this, either by.

- **key transport**—one party generates and distributes a secret key
- **key agreement**—both parties jointly agree on a secret key.

3.5.1 Diffie-Helman (DH) Key Agreement

A popular example is the **Diffie-Helman (DH)** key agreement method which was published by Whitfield Diffie and Martin Hellman back in 1976, The DH key exchange scheme allows two parties to jointly establish a shared secret key over an insecure channel. This key can then be used to encrypt subsequent communications using a symmetric-key cipher in TLS protocol, see Sect. 3.6.1. In Fig. 3.7, a DH handshake for a sample secret key agreement scenario is illustrated. Circled numbers in picture are referring to numbers in brackets in text. For a sample calculation of a joint pre-master secret **S**, parameters with the following values have been used:

- modulus $\mathbf{p} = 25$
- base $\mathbf{g} = 38$
- client: secret random $\mathbf{a} = 2$
- server: secret random $\mathbf{b} = 5$

On client request, the server starts the key agreement process. The protocol uses the multiplicative group of integers modulo **p** (aka "modulus") where **p** is prime, and **g** is a primitive root of prime number **p** (aka "base"). Integers **g** and **p** are random numbers, but carefully selected as seed parameters for the calculation process and will be shared with the client (1).

The server will then generate a random number **b**, and based on previously generated values for **g** and **p**, the server will then generate $\mathbf{B} = \mathbf{g^b}$ **mod p**. On client side, same operation is done with secret random a: $\mathbf{A} = \mathbf{g^a}$ **mod p**. Both results **A** and **B** are shared with the other party (2).

Now, the trick is that on both sides the same secret $\mathbf{S} = \mathbf{g^{ab}}$ **mod p** can be calculated:

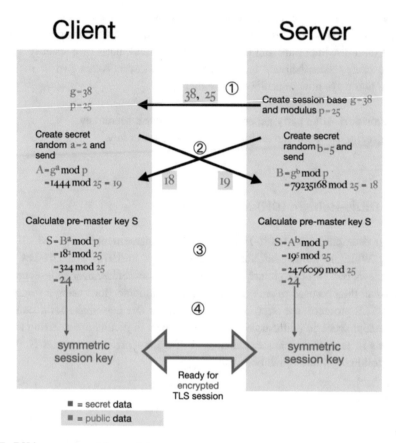

Fig. 3.7 DH key agreement (example)

$$S = A^{b} \bmod p = (g^{a} \bmod p)^{b} = g^{ab} \bmod p.$$
$$S = B^{a} \bmod p = (g^{b} \bmod p)^{a} = g^{ba} \bmod p.$$

Only a and b are secret, all other values are sent in clear. $S = g^{ab} \bmod p$ is called the pre-master secret (3).

Here, for demonstration purposes, in this example only small numbers have been used. With **p** = 25 only 25 possible results of a pre-master secret are possible (n mod 25) and can be determined shortly. However, if **p** is a large prime of—let's say—1000 bits, then even the fastest known algorithm on fastest computer cannot find random number **a** based on given public values of **g**, **p** and (**g^a mod p**) and (**g^b mod p**).

Based on this shared pre-master secret **S**, both client and server can generate a master key (4). This nonce will feed a pseudo-random function (PRF) function which will be used to produce a securely generated output of arbitrary length. For use in conjunction with the TLS protocol, for example, a length of 256 bit would be selected. This output number will then be used as a temporary secret 256-bit AES session key on both sides.

3.6 Secure Data Transmission

By nature, Internet connectivity and data transmission are important IoT ingredients. Independent of used network technology, the Internet protocol suite (aka TCP/IP) or the OSI model [6] are specifying two layers which are commonly used for data transfers between an IoT device and the IoT server: **Transport Layer** and **Application Layer**. On Transport Layer, TCP or UDP can be used for simple, but efficient communication. Application layer protocols are used for direct user interaction with the software application and often associated with particular client–server applications, e.g., HTTP for the World Wide Web, Telnet for an interactive text-oriented virtual terminal connection or FTP for file transfer. For an IoT application, protocol selection will depend on specific requirements. For some, HTTP might be a good choice. For IoT applications with devices either publishing or subscribing to IoT data packets, MQTT (Message Queueing Telemetry Transport) might offer appropriate support. For secure data transmission, TLS has gained popularity in the IT world because it combines various cryptographic functions such as a public-key methods for user authentication, MACs for data integrity and symmetric cipher for data encryption.

In order to ensure bullet-proof communication and remote control of IoT devices, IoT projects will have to implement an appropriate level of security and protection against potential attacks or attempts to misuse the IoT application. Independent from technology used for data transmission, a secure channel between communication partners will have to protect integrity and confidentiality of data. This end-to-end security will ensure that nobody can understand or modify data transferred from one endpoint to another (typically from an IoT device to a dedicated IoT server or vice versa). Typically, an IoT server is located in a safe environment. But for many IoT use cases, end-to-end security requirement is particularly challenging because involved IoT devices are mobile and/or unattended, i.e., exposed to risk.

For use in IoT devices, all vendors of cellular network modules are bundling their products with firmware-based software stacks offering features or options to support TLS. For an IoT device, in most cases the endpoint of a secure TLS channel will be an integrated "secure element" inside the network module. For a software-only implementation, the TLS/SSL software stack will be part of the module firmware and accompanied with a dedicated set of AT commands for use by the IoT application. Key storage and crypto operations will be performed by the module MCU. See Sect. 4.3.7 for further information.

3.6.1 TLS Protocol

For preparation of a TLS-based secure communication session, IoT client and server will have to authenticate each other, to negotiate some protocol parameters and to share a data encryption key.

Key parameter is the **cipher suite** to be used for exchanging messages. A cipher suite specifies used cryptographic algorithms and key lengths. For example, TLS_DHE_RSA_WITH_AES_256_CBC_SHA means:

- tunnel type: TLS
- public-key algorithm for digital signatures and PKI: RSA
- key exchange method: DHE (Ephemeral Diffie-Hellman)
- symmetric algorithm for data encryption: 256-bit AES with CBC
- hashing method: SHA

Figure 3.8 outlines the TLS handshake between both parties indicating various functional groups of tasks performed during this exchange of datagrams needed to up a TLS session. As a start, the IoT device (client) initiates the handshake by sending a "hello" message to the server. The message will include the used TLS version and cipher suites supported by the IoT device. IoT Server will respond with a "hello" message including the connection parameters it has selected from provided list. Certificates will be exchanged between both parties allowing authentication using public-key cryptography, typically based on RSA cipher. After successful authentication the agreement process for the TLS session key will start, details are explained in Sect. 3.5. When done, both parties will have created a temporary symmetric key for data encryption according to the agreed cipher suite. Finally, both client and server are sending their "finished" messages which are encrypted using the new session key—indicating that a trusted connection channel has been established. At this point, first IoT application data can be sent.

As a default setting, TLS is focusing on authentication of the IoT server, i.e., the client verifies the identity of the server. This might be fine for traditional browser-server interaction, but for IoT applications also user devices are potential candidates for hacking and identity theft. As a consequence, mutual authentication is required to cover IoT devices

Fig. 3.8 TLS handshake

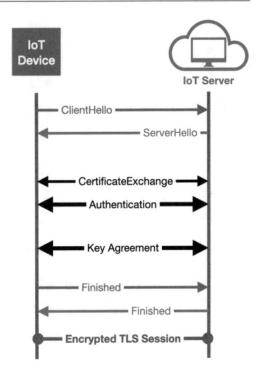

as well and extend overall security scope. **Mutual TLS (mTLS)** is an option of the TLS protocol [7] which can be used for this purpose. With mTLS, a two-way authentication of both parties is performed.

References

1. Hopcroft, J. E., Motwani, R., & Ullman, J. D. (2007). *Introduction to Automata theory, languages, and computation.* Addison Wesley.
2. Paar, C., & Pelzl, J. (2011). *Understanding cryptography: A textbook for students and practitioners.* Springer.
3. OpenSSL. Retrieved 22 March, 2022, from https://www.openssl.org.
4. OpenSSL PKI Tutorial. Simple PKI. Retrieved 22 March, 2022, from https://pki-tutorial.readthedocs.io/en/latest/simple/.
5. International Telecommunication Union. X.509: Information technology—Open Systems Interconnection—The Directory: Public-key and attribute certificate frameworks. Retrieved 22 March, 2022, from https://www.itu.int/rec/T-REC-X.509.
6. International Telecommunication Union. X.225 : Information technology—Open Systems Interconnection—Connection-oriented Session protocol: Protocol specification. Retrieved 23 March, 2022, from https://www.itu.int/rec/T-REC-X.225/en.
7. IETF. "The Transport Layer Security (TLS) Protocol Version 1.2". Retrieved 23 March, 2022, from https://datatracker.ietf.org/doc/html/rfc5246#section-7.4.6.

Ingredients for Secure Design

4

Risk assessment will determine case-by-case which threats are expected and to identify applicable attacker types and their potential, etc. in order to specify an appropriate set of countermeasures to be implemented at a certain protection level (see Chap. 2). Typical attacks are targeting the electronic system of an IoT device ("electronic intrusion") in an effort to change its behavior or modify transmitted IoT data. As a consequence, the requirement specification of each IoT device will ask the designer to look for suitable hard- and software elements to be used as countermeasures. IoT security needs are tailored to meet specific use cases and applications, but many common soft- and hardware ingredients are available as **configurable standard products** and can be used for most IoT applications.

4.1   Intrusion Protection/Detection

As explained in Sect. 2.6, typical attempts against IoT device resources are **remote attacks**, i.e., will be applied via network connection. In fact, many device locations will not allow attackers to open the case of an IoT device in order to get access to internal resources and perform local attacks using tools and external equipment. But some IoT use cases are aiming at unattended or at least at uncontrolled environments where attackers can spend plenty of time to set up a **local attack**. Designers of IoT devices facing risk of local attacks will have to put appropriate countermeasures in place.

Another serious physical threat scenario has been described in Sect. 2.4. If an attacker is able to obtain a sample IoT device, he can perform **in-depth local hard- and software analysis** in order to prepare an attack to be performed later. This approach allows a cyber-criminal to perform time-consuming investigations and elaborate a sophisticated attack or an attack template to be provided to less experienced "type-1" attackers.

© The Author(s), under exclusive license to Springer Nature Switzerland AG 2022
K. Heins, *Trusted Cellular IoT Devices*, Synthesis Lectures on Engineering, Science, and Technology, https://doi.org/10.1007/978-3-031-19663-8_4

As a first level of protection, the IoT device case should be **tamper-resistant** in an effort to complicate opening it. But skilled attackers will identify more sophisticated intrusion methods and/or find a way how to apply pentesting attempts. In fact, there are several ways how to prevent physical intrusion. Besides putting technical countermeasures in place, applied IoT deployment policy can **restrict device ownership and access to product documentation** by asking customers to identify and to register before they can take part (see Sect. 2.7 against type-3 and type-4 attackers). Another approach is to rent devices to users instead of selling them. This means that an end-customer will have to **return a used IoT device** at some point of time, e.g., after contract termination. In this case, a **tamper-evident packaging** will probably help because it indicates that it has been opened nonreversible manner. This will probably stop intruders because penalty payment and/or legal consequences would apply after return of the used IoT device.

But mechanical tamper-evidence does not work against any less obvious attacks, i.e., attacks which do not leave visible traces. For second-level intrusions more sophisticated approaches are required for **tamper detection**, i.e., subsystems to inform the operator about non-evident tamper attempts. Other protection approaches are providing **tamper responsiveness** which are pro-actively engaging countermeasures (e.g., deletion of stored data) whenever an attack has been identified.

In general, intrusion protection approaches are aiming at increased resistance of an IoT device against remote or local attempts to access internal resources. They can be categorized like this and will be discussed in subsequent paragraphs:

1. Simple mechanical countermeasures (tamper evidence, tamper resistance)
2. Countermeasures against side-channel (SCA) and fault-injection (FI) attacks
3. Active sensing
   a. physical
      i. monitoring (tamper detection)
      ii. immediate action (tamper responsiveness)
   b. remote (network)
      i. monitoring (tamper detection)
      ii. immediate action (tamper responsiveness)
4. Sign-of-life indication (incl. watchdog).

4.1.1 Mechanical Intrusion

Simple mechanical means cannot stop well-educated attackers to perform "offline" vulnerability analysis but might help whenever an IoT device is located in a public but untrusted area, e.g., in a taxi or sports arena or museum. **Local attacks** will require access to the IoT device interior, typically to internal connectors or leads (pins) of electronic components. In attended environments, an attacker cannot spend much time to apply a pre-prepared local attack and will avoid attracting public attention. In order to **demotivate potential attackers** to perform a local attack, an IoT device can limit risks of physical intrusion if a tamper-resistant case is used.

For first-level protection, IoT product owner might consider using **tamper-resistant** case fasteners, i.e., **security screws** which deter or prevent disassembly, opening or local intrusion into the device. This is a low-cost approach particularly for IoT device deployments which are accessible to the public and where you cannot prevent physical tampering attempts. Security screws are distinguished by having an unconventional drive, making tampering with the screw more difficult, if not impossible without the matching driver. A special approach is called **one-way screws** (or irreversible screws) because of their drive style, see Fig. 4.1. The head of the screw features a slotted drive that is designed in a way that the screwdriver can be turned in one direction only. This is achieved by manufacturing the drive-in quadrants that are gradually raised to accept the driver bit when turning the screw right and reject it when turning the screw left. This makes installation of the screws easy, requiring only a standard slotted bit, but removal of the screw difficult (if not impossible) without the corresponding bit or a specifically designed removal tool. Today, security screws are often used for license plates, equipment fittings in schools, bathrooms or in transportation vehicles.

Fig. 4.1   One-way screw

Fig. 4.2   Security seal

Security seals can be used to deter and to detect unauthorized opening of a closure. A seal will deter users of an IoT device who are not owning it because violation of the seal will indicate tampering of the device and cause trouble after return. Indicative seals are the most commonly used type and are found in countless applications. Unlike locks, seals are closed only once. They are easily broken or cut open with common tools or sometimes by hand but serve the function of providing **tamper evidence**. Simple seals use plastic bodies, but also metal versions with a cable are available (see Fig. 4.2). Adhesive labels and tapes can also be used.

4.1.2   SCA and FI Attacks

This kind of attacks typically require physical intrusion of the IoT device case because the attacker is using equipment to apply faults and to **measure and analyze information leakages**. This means that methods to increase case tamper-resistance or to implement reliable tamper-evidence will have to be applied in the first place. Section 2.4.2) already provided an idea of various attack types and sophistication of **side-channel attacks (SCA)** and **fault-injection (FI) attacks**, which are requiring experts on both sides. But interestingly, e.g., for side-channel power analysis, tools are available for sale to everybody. Target audience are manufacturers who are planning to implement efficient countermeasures against. But for sure, they can also be used as tutorials for cybercriminals. Some side-channel attacks require technical knowledge of the internal operation of the system, although others such as differential power analysis (DPA) are effective as **black-box attacks**.

Some countermeasures are easy to implement. For example, since side-channel attacks are trying to exploit internal secrets, as a countermeasure, usage period for keys should be limited, i.e., they should regularly expire and get replaced. Alternatively, random session key should be used, if possible. For FI attacks, since a fault may be non-intentional, countermeasures are the same as the one used for safety: redundancy, error detection and monitoring. Hardware countermeasures are adding resistance, but corresponding **secure programming** techniques can help to protect against operational faults including attack-generated faults. This is common practice for automotive products, and corresponding tutorials can be used by device designers to educate themselves.

Alternatively, designers can work with consultants to implement countermeasures against common SCA and FI attacks. For this purpose, security evaluation labs are good starting points, see Table 2.8 which is providing a worldwide list of candidates with embedded security expertise which are either offering design consultancy themselves or otherwise will be able to refer to suitable service partners.

In general, for designs requiring type-3 or type-4 resistance (see Sect. 2.7) utilization of **certified components** is recommended (see Sect. 2.10). In particular, smartcard modules or so-called secure elements (see Sect. 4.3.5) have been designed to generate high-quality random numbers and keys, store them, and perform cryptographic operations in a tamper-proof environment, i.e., on a single secure microcontroller chip. During the evaluation of a security component, SCA and FI attacks are standard pentesting instruments and used by the test lab to verify the quality of implemented protection mechanisms. As a consequence, the security certificate of a component will provide evidence of resistance against known SCA and FI attacks.

4.1.3 Active Intrusion Sensing

On top of mechanical protection, active sensing of case-internal parameters can be used as a second-level protection against mechanical intrusion. Typically, switches or vibration sensors can be used for this purpose—or a light sensor. The idea is to detect obvious tampering or suspicious activities for later use (**tamper detection**) or immediate action (tamper-responsiveness), [1] is providing some background information. The IoT device can keep identified events and associated parameters internally in a **non-volatile log file** which can be used later as evidence or for investigation purposes. In fact, this feature been listed as a standard requirement in Sect. 2.5. Besides mechanical intrusion events, other parameters like supply voltage, clock frequency, temperature variations, etc. might be relevant for detecting and monitoring of abnormal operating conditions. Of course, active sensing and tamper-detection mechanisms **work with battery powered devices only**. For illustration, Fig. 4.3 is showing involved elements of two different sample tamper scenarios which are described here:

1. **Violent case opening.** Incident light will be detected by an internal digital sensor with adjustable threshold which is used as evidence that somebody has started to open the case. This event will trigger an interrupt of the host MCU leading to immediate action. Corresponding interrupt handler will record the event in non-volatile memory (tamper-detection). In addition, and as a carefully pre-defined response to this kind of detected event, the interrupt handler can **delete critical assets** (e.g., cryptographic keys, payload data) or any critical part of the embedded device software (tamper-responsiveness). Instead, or in addition, the device MCU can submit an **external alert** immediately after detection of a mechanical intrusion, e.g., via wireless network. Refer to Sect. 4.3.8 for more information about suitable sensors.

2. **Manipulation of supply power.** Forced power glitches or supply voltage beyond specified operating conditions might indicate an FI attack and should be detected and recorded. For monitoring purposes, an analog–digital-converter (ADC) input of the MCU can be used, if available. Refer to Sect. 4.3.3 for more information. For later verification purposes, the MCU can track supply voltage and log values.

Fig. 4.3 Active intrusion
sensing

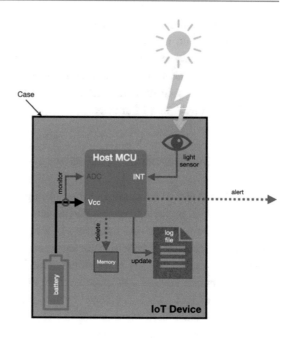

4.1.4   Network Attacks

For **remote IoT intrusion** attempts, attackers can use the network connection of the IoT
device. Attackers might be able to prepare and explore all options (see Sect. 2.4). But, in
any case, available functions for network access to device-internal resources depend on
implementation and should be limited by design. A fundamental security recommendation is to minimize potential entry doors for intrusion, so each IoT device should offer
a carefully selected set of external services only—and disable all other options allowing
users to access device data or configuration, if not required. In addition, **only adminis-
trators** should have access to essential device functions, e.g., to manage a secure device
firmware update. In an effort to limit vulnerability against "elevation of privilege ("E")
attacks" (refer to Sect. 2.3), no other IoT users (i.e., non-admins) with alternate user profiles should be allowed. Instead of providing access to raw IoT device data, IoT operators
should offer an extra web portal to provide consolidated and dedicated user data. On top
of this, **strong user authentication** (i.e., using tamper-proof identities) should be used to
rule out any unauthorized access to implemented device functions.

For most IoT device designs, the built-in software stack of an off-the-shelf modem
will handle all low-level handshakes and data exchanges with a communication partner.
In this case, an external party will first have to connect to a **network socket** before any
further internal data exchange will be granted by the module. This is a prerequisite for
external parties to use application-layer services like FTP or Telnet, but also applies to
simple data exchanges via TCP/IP protocol. See Fig. 4.4.

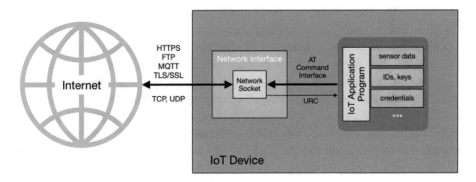

Fig. 4.4 Network socket

A network socket is a software structure within a network node that serves as an end-point for sending and receiving data across the network. It has to be specified for each communication scenario with an IoT device. For this purpose, the host MCU resp. embedded IoT application will have to set up a corresponding **PDP context** (PDP = Packet Data Protocol) and activate it prior to any transmission activity. Details of this process are found in manufacturer manuals, for example, by Quectel as a TCP/IP application note for their BG95 module [2]. Besides other parameters, each PDP context in combination with a socket service will define

- the individual device IP address and
- an authentication method, either PAP (username, password) or CHAP (challenge—response).

In addition, different data access modes are available: buffered, direct push, transparent. In buffered mode, if the module has received the data from the Internet, it will keep the data and report a URC (unsolicited result code) to the host MCU (see Fig. 4.4) which can read the data via AT command, verify it and decide what to do with it. This access mode avoids risk of pushing malicious data from an external source directly into the device. Finally, a carefully defined user concept and a restricted socket configuration incl. activated authentication will complicate remote attacks and will block attempts of external network node to access device assets.

For complex multi-user IoT devices, protection against network intrusion will be more difficult to implement. In fact, for networked IT systems (servers, etc.) it is common practice to put an **intrusion detection system (IDS)** in place. An IDS is a device or software which is monitoring access to a computer system and able to recognize unauthorized or malicious activities or a suspicious behavior of a system. By nature, IoT devices are connected to the Internet and vulnerable against harmful access in a similar way. Considering increasing utilization of IoT applications in many areas, also for mission-critical purposes, business owners will have to take a close look at this particular attack entry path. But IoT devices are not based on common platforms or standards. Instead, most IoT devices have been designed for a specific target IoT use case. They are based on different system architectures, use different operating systems, etc. On top of this, storage capacity (e.g., for log files) of typical IoT devices as well as computational power (for analysis or detected events and associated parameters) are limited. Consequently, today there are no commercial off-the-shelf IoT IDS products available. But intrusion detection techniques originated from existing IDS products are a good starting point for IoT-specific IDS implementations.

Different intrusion detection methods are leading to two main IDS sub-categories: Signature-based Intrusion Detection System (SIDS) and Anomaly-based Intrusion Detection System (AIDS), see [3].

- A **SIDS** is based on sequences of commands or actions which have previously been used for attacks. The idea is to extract significant activity patterns and create an **intrusion signature** database. A SIDS is monitoring current activities or log files, applies pattern matching methods to identify similarities with known intrusions and finally will submit an alert if a match is found.
- An **AIDS** is based on a model describing a normal **user behavior** of a computer system. The idea assumes that an attacker behavior differentiates from the standard behavior. Any significant deviation between an observed user behavior and this model is regarded as an anomaly which is interpreted as an intrusion and trigger an alert. During a training phase, the AIDS will have to learn from normal traffic and build a model of standard user behaviour.

Most IoT applications require tailored IDS solutions for their IoT devices because different platforms are used, and no "standard" user behavior can be defined. But both SIDS and AIDS intrusion detection methods might inspire IoT manufacturers to utilize intrusion signatures and/or specify a normal user behavior. Corresponding pattern matches can be used to support device-specific tamper protection mechanisms.

4.2 Standard Embedded Security Functions

Besides countermeasures against intrusion attempts, other embedded functions are addressing common security requirements of an IoT device. Some are optional or might be required for advanced type 3 or type 4 resistance (see Sect. 2.7) only, but many versatile embedded security ingredients are included or bundled with standard components which are required anyway (e.g., a watchdog timer). Other security functions are offered by hardware manufacturers of network interface modules (e.g., generation of random numbers and cryptographic keys) or host MCUs (e.g., a secure boot feature) as ready-to-use software modules and can implemented easily. In fact, tailored secure IoT ecosystems can be built with customizable off-the-shelf building blocks and (cloud) services in an efficient way.

4.2.1 Sign-Of-Life Indication and Watchdog

Typical IoT threats are aiming at identity spoofing or manipulating IoT devices (refer to Sect. 2.3). Alternatively, a tailored software intrusion or denial-of-service (DoS) attack or simple physical destruction (incl. vandalism) will **cause device unavailability and destroy a corresponding IoT business model** at the same time. This kind of attacks are causing serious damage and financial loss. Countermeasures might help and internal intrusion detection mechanisms can create internal log files containing valuable information for later investigation. But as a first level of tamper-evidence for deployed IoT devices in the field, the operator needs to know that each of them is working, i.e., alive. For this purpose, the IoT device should be able to submit a sign-of-life message to the operator and provide evidence that it is still operational. There are **two approaches** for implementation:

- **push**: scheduled or
- **pull**: on request.

Characteristics of each IoT use case will determine, which approach is most appropriate. For IoT devices which are permanently interacting anyway, there is no specific action required because device operation is evident. But careful consideration is needed for battery-powered IoT devices which are designed to minimize energy-consuming transmission periods which are dramatically decreasing battery lifetime. By nature, this kind of devices are not permanently online and might report regularly, but infrequently (e.g., an environmental sensor). Others might not be designed for scheduled transmission of IoT data at all, i.e., they submit an alert message only in case of a predefined event (e.g., a fault detector). Infrequent periodic transmission events might not provide sufficient evidence either, so IoT operators might ask for an option for triggering an IoT device to transmit a suitable sign-of-life message.

Figure 4.5 illustrates both approaches. Periodic push transmissions are triggered by an MCU-integrated timer or another reliable and configurable source, e.g., a cellular network timer.

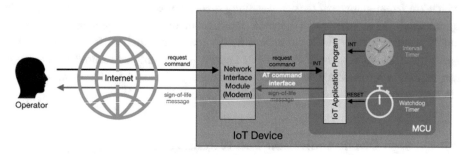

Fig. 4.5 Triggers for sign-of-life message

For **operator request for a sign-of-life** from an IoT device, a simple PING command can be used. PING is a command-line utility, available on virtually any operating system with network connectivity, that acts as a test to see if a networked device is reachable. Pinging the IP address of an operational IoT device means to send an echo request to the network socket which will return an echo reply. This response provides evidence, that the device is powered up and its network interface is working. But it does not tell you that the IoT device is working properly. In particular, the PING command does not involve the host MCU and does not verify the status of the IoT device system hard- and software (see further explanation in Sect. 4.2.3).

For a more comprehensive, meaningful sign-of-life message the IoT application software will have to be prepared accordingly so that operator can send a dedicated well-defined control word to the device. This piece of data will be interpreted by the host MCU as a software interrupt and will trigger the execution of a dedicated handler. Then, this routine can perform some internal tests and collect some meaningful **status data to be consolidated by the IoT application as a sign-of-life message** and finally to be transmitted to the operator's IP address or by SMS. Based on this approach, an IoT device will be prepared to return an application-specific status—on request of an authorized external user. As a consequence, the operator can take appropriate action.

A DoS attack or an unintended side-effect of an intrusion attempt might lead to another, completely different fault scenario where the device is still seems to work, but differently as expected or software execution has crashed or resides in a deadlock state. In some cases, a remote reboot command might solve the problem, but might not work if an MCU has gone out of control. In any case, an **onboard auto-recovery function** might help in the first place. For this purpose, a **watchdog timer** can be used to catch the microcontroller and return execution of the embedded IoT application to a pre-defined entry point. A watchdog timer is an integrated simple countdown timer which is used to reset a microprocessor after a specific interval of time (see Fig. 4.5). Then, the boot process will return to normal execution with a well-defined sequence of instructions and restart the watchdog timer. After each restart, the watchdog will begin timing another predetermined interval. As a consequence, utilization of a watchdog timer significantly supports reliable operation of any deployed IoT device.

4.2.2   Secure Firmware Update

This function is used to revise the embedded IoT application program. It is a standard security feature which is required for IoT ecosystems using cryptography and related credentials which are at risk during the whole product lifecycle. Security updates will refresh countermeasures and ensure to keep device protection in place at the same level over time. In addition, an updated IoT application software can correct bugs, add new features, and optimize overall device operation (refer to Sect. 2.5). But it is also an option for keeping an IoT device in place after a successful attack, e.g., if essential assets have been modified. In this case, a secure firmware update avoids device exclusion resp. its replacement.

Fundamentally, a firmware update is an image file, i.e., a one-to-one copy of the memory section which containing the IoT application software. The update process is supposed to replace the existing image with a new revision of this image. For this purpose, the IoT operator (or an authorized entity) will have to transmit this file to the device. For cellular user devices this process is also called FOTA which stands for "firmware-over-the-air". The Internet protocol suite defines a standard application-layer protocol for file transfers called FTP (File Transfer Protocol) which can be used with any commercial cellular network interface module (see Sect. 4.3.7). But in order to protect the FOTA process against tampering, an additional security layer has to ensure authenticity, integrity, and confidentiality of the firmware image before it can be activated. Typical solutions are hybrid processes based on **signed and encrypted firmware image files** which are verified and decrypted on client side. File signature and verification follows the process described in Sect. 3.2, it is illustrated in Fig. 4.6. After reception of the firmware image, the IoT device will the verify the signature of the sender as well as the integrity of the file.

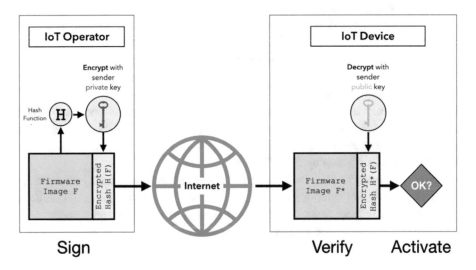

Fig. 4.6 Signed firmware update process

This is a good start, but not sufficient yet because—as explained before in Sect. 2.4.1—the IoT application program contains sensitive information which should not be disclosed to anybody. Consequently, the file should use an encrypted transmission channel. For this purpose, a standard TLS/SSL connection can be used on top of mentioned FTP protocol. Both protocols are handled by most off-the-shelf network interface modules. On top of this, comprehensive FOTA management and provisioning solutions are being offered, see Sect. 4.3.7.

4.2.3   Integrity Check and Secure Boot

If all tamper protection and intrusion detection measures have failed, an attacker has been able to modify the embedded software or configuration of critical hardware components. This might have been done in a way that nobody will notice any change of expected IoT device operation. Therefore, the IoT device should verify its own system integrity itself and make sure that relevant details have not been changed. For this purpose, a dedicated embedded software module will have to "measure" all relevant **functional determinants of the device**, this process is called "integrity measurement". Besides the IoT application program itself, also the configuration settings of the modem (network interface module), sensors, actuators, and the host MCU itself are representing all details of the IoT device operation. In addition, essential assets like PKI certificates or credentials used for authentication should be included. **Integrity measurement** will have to retrieve component configuration data as well as relevant parts of the embedded software and create a compact "fingerprint" reflecting the actual device functionality. Typically, a hash function is used for this purpose (see Sect. 3.2 for further information). The desired condition of an IoT device is measured during its initialization (or after a firmware update) by an authorized person, e.g., during device deployment. This initial measurement M_0 will be used as a reference ("Integrity reference M_0") during all subsequent integrity checks (see Fig. 4.7). Of course, reference value M_0 must be securely stored in a tamper-proof way (e.g., in an OTP memory location). Same applies for the credentials used for authentication of the person who is allowed to put a new measurement M_0 value in place.

During an **integrity check**, the same measurement procedure will be used to create M_n which then stands for the actual device configuration. Only **if M_n equals M_0, the integrity check was successful**. A negative result should trigger an alert message by the device for immediate action by the IoT operator.

Typically, an integrity check will be part of a **secure boot** process, i.e., after power-up the IoT device will not enter operational state without a successful integrity check. This will be sufficient in most cases, but additional or even periodical integrity checks can be performed at any time during field operation, if needed.

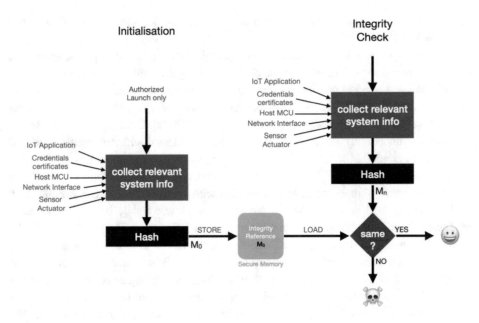

Fig. 4.7 System integrity check

Complete off-the-shelf secure boot solutions are offered as an add-on by several manufacturers of key IoT components. A versatile concept how to implement comprehensive device integrity verification assisted by a secure element (SE) is provided later in this book, see Sect. 5.2.3.3.

4.2.4   Key Generation and Provisioning of Device Credentials

For IoT ecosystems using public-key cryptography to protect network communication, key generation and key deployment is a major topic (recall Sect. 3.3 for background information about Public-Key-Infrastructure and certificates). For each IoT device, an immutable and unclonable identity is required for **onboarding an IoT device**, i.e., to authentice a device within the IoT network and to establish a secure communication channel with this device. There are several options how and where to **create a key pair for use in a PKI**, but first of all and as a starting point, a unique identifier is required which is inseparably coupled with the device hardware. Ultimately, this requirement is met by so-called **physical unclonable functions (PUF)**. PUFs are most often based on unique physical variations which occur naturally during semiconductor manufacturing (see [4]). Consequently, they are used by chip manufacturers to address security applications for identification, eCommerce, etc. As an alternative to a PUF-based key as a device identity, the semiconductor manufacturer can also manage to create and to **inject a unique identifier into each chip** during production.

On top a unique ID, many secure cryptographic mechanisms and protocols require good random numbers. Therefore, a **random number generator (RNG)** used in crypto-graphic products need to provide random and unpredictable data. Hardware-based RNGs are also called "true" RNGs (TRNGs) because created from a physical process instead of a software algorithm known as "pseudo" RNGs (PRNGs). Term "entropy" is used to measure the uncertainty of the RNG data output leading to perfect unpredictability of a new cryptographic key. A TRNG can deliver high-quality random numbers which are required for **generation of PKI key** pairs—to be created either inside the IoT device or injected from an external source. In combination with a PKI certificate, this **certified IoT device identity** will be trusted within the IoT network and finally be used as a **Root-of-Trust** which is anchored in the hardware of each IoT device. In order to certify the identity of an IoT device (i.e., a PKI certificate), the device identifier (plus some extra metadata) and associated public portion of the key pair will be signed by a **trusted-third party, a so-called CA (Certification Authority)**. This certificate can be stored inside the device which will deploy it to communication partners, or it can be published within the IoT community, as needed. Figure 4.8 is an idealized illustration how to create a PKI certificate for an IoT device.

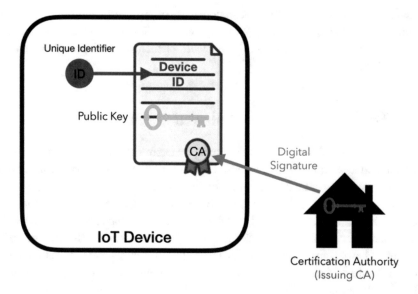

Fig. 4.8 IoT device certificate

For new PKI-based IoT projects, generation and maintenance of device credentials is a fundamental and critical task requiring careful consideration. In particular many options are available for the generation of new PKI keys and issuing certificates for them. On top of this task which is required to onboard new devices to the IoT network, also the maintenance of stored credentials during the product lifecycle is crucial. Practically, this means to decommission selected devices, revoke certificates, and delete credentials inside the device remotely. Starting point for related considerations incl. selection of PKI partners is the hardware location where to store sensitive device credentials—and where to process them. In an effort to increase IoT security, and in order to reduce vulnerabilities and data leakages (refer to Sect. 2.4) it makes sense to use an **integrated secure environment for credential storage and processing**. Consequently, a **secure MCU is the most appropriate place where to generate or inject and keep IoT device credentials** (see Sects. 4.3.5 and 4.3.6 for further explanation and design options).

In general, secure processes are required for manufacturing and delivery of security ICs anyway, and semiconductor manufacturers are leveraging this qualification for versatile **provisioning services** for cryptographic keys and/or certifications. Some of them are even offering PKI services themselves. Or customers can use their own PKI or one the major third-party IoT clouds from AWS, Google, etc. Typically, for large IoT installations there will be more than one CA, and the issuing CA (e.g., the MCU manufacturer) will be a lower-level part of a PKI hierarchy which has been applied for a particular project. The top-level CA (aka "root CA") will be selected by the business owner of an IoT project, and acts as a trust anchor. For example, for governmental projects like a national smart-meter typically a federal agency replacement will be selected as a root CA. The idea is that the assigned **root CA delegates the permission to certify IoT devices** "on behalf" of the owner and establish a **hierarchical CA structure**. For this purpose, the root CA (level #1) will authorize next level #2 CAs and certify their public key of so that certificates issued by a level #2 CA can be verified as valid within by participating parties with the IoT network. Verification of a device certificate is performed level-by-level by walking through this **chain-of-trust** up the root CA. More levels can be added, if needed. This principle is illustrated in Fig. 4.9 with 3 PKI levels as an example and device certificates issued by a level #3 CA.

For larger quantities (e.g., starting from 100 k units) manufacturers are also offering **customized** provisioning services allowing secure on-site injection of die-individual keys and certificates. But a pre-provisioned simple configuration with default keys and certificates can be used as a start for many use cases, also for onboarding to an IoT cloud. Later, new credentials or certificates can be loaded remotely and, for example, replace expired keys. In combination with secure firmware uploads, **remote in-field provisioning** services are increasing product lifetime of deployed IoT devices by refreshing their software and maintaining the intended strength of implemented IoT security measures.

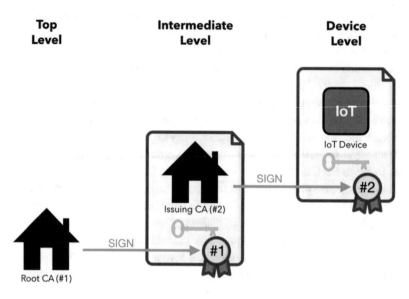

Fig. 4.9 Certificate hierarchy

4.3   IoT Design with Off-The-Shelf Elements

Good news for IoT project owners is that the design of a (secure) IoT device has never been easier because many essential hardware building blocks and other ingredients (driver software, communication protocol stacks, IoT clouds, etc.) are offered as **universal products and services**. Instead of developing a custom product which are dedicated to address the customer application, standard products are offering a superset of required features which are configurable according to customer needs as much as possible. Consequently, the IoT application designer will not have to develop everything from scratch. Even if a standard product might not be the perfectly tailored solution for every IoT project, it will reduce development effort, speed up time-to-market and reduce overall cost. From a customer perspective, the **standard-product-based design approach** particularly pays off for low-to-medium-scale projects with a demand of less than 100 k units p.a.

Standard hardware products are so-called ASSPs. What is an ASSP? An ASSP stands for Application-Specific Standard Product. It is an active chip component which has been designed to address a specific electronic application (e.g., a cellular network interface). In contrary to customer-specific ASICs (ASIC = Application-Specific Integrated Circuit) which is owned by the user of this chip, an ASSP is designed to be used by different customers and offered for sale by chip manufacturers. The ASSP business model is based on the business assumption that the addressed target application will be an appealing market for **many customers**—leading to high volumes and sales numbers.

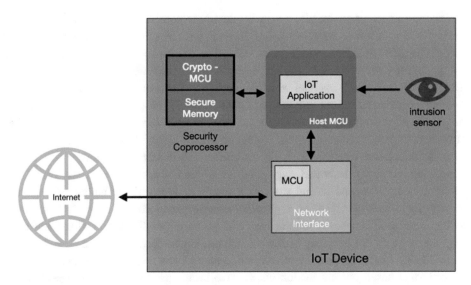

Fig. 4.10 ASSPs for secure cellular IoT devices

For the chip manufacturer, a fundamental success enabler is a powerful external sales channel allowing them to focus on design, documentation, online support, product marketing, but to delegate direct customer interaction. Typical ASSPs for use in secure cellular IoT device designs are illustrated in Fig. 4.10. For secure cellular IoT projects, ASSPs and services are available for these categories which are handled in the following chapters:

- Host Microcontrollers and Security Coprocessors
- Cellular Network Interface Modules
- Intrusion Sensors.

The following sections are providing a market snapshot of ASSP candidates for cellular IoT device design and relevant selection criteria—at the time of writing.

4.3.1   Hardware Suppliers and Online Support

Good news for IoT application developers is that the "Internet of Things" has been elected by all market players as the #1 top priority application for the IT electronics business. All contributors like semiconductor manufacturers, network operators, distributors, IT service providers, etc. are trying to benefit from promising IoT market outlook and take their share. For IoT device designers this means that they can expect to receive a decent level of support for their engineering work.

On top of this, most manufacturers of electronic components and subsystems have learned how to handle a large number of different customer projects via sales partners or online channels. In fact, most chips are offered as standard products accompanied by a **comprehensive set of documentation**, evaluation tools and a design kit. Objective is to provide self-explanatory material which is supposed to answer most questions in order to minimize customer need for one-to-one support. All relevant product information should be **published online and downloadable** via manufacturer website. Usually, it also allows customers to order product samples, evaluation boards, design kits, etc. directly.

In addition, and whenever needed, an **authorized dealer (distributor)** will be the day-to-day business partner and entry-point for all kind of customer requests. Traditional distributors are independent and work as a supply partner offering additional customer services incl. stock management, application expertise and technical support. As an alternative, commercial customer requests can also go to **online distributors** like Mouser or Digi-Key who do not offer any additional support, but competitive prices.

IoT design engineers are particularly online-minded and might take decisions to select a component based on information which have been extracted from online sources. In general, manufacturer websites are most important **self-service repositories** for product information and a common starting point for application designers to prepare for competitive product comparisons. Besides technical information like datasheets and user manuals also white papers, presentations and videos are available for download. Design kits should include drivers, sample source code, schematics, guidelines for PCB layout, etc. For components like cellular network modules or sensors which are specifically addressing devices for IoT applications, many manufacturers are offering **IoT-specific application notes and design tips** in order to support implementation and to speed up customer time-to-market. In particular, they should explain how to perform application-specific adjustments, e.g., which features have been implemented to save power consumption and/or how to configure a NB-IoT network cell according to application requirements.

On top of product information, manufacturers should offer **interactive support services**. A popular online support instrument is a virtual community where people with a particular common interest meet online and exchange information. For this purpose, manufacturers of electronic components offer a community platform with discussion boards for product-related topics. These are places where users can ask questions and share material with other community members. **Communities** are managed by a company moderator and supported by product experts, but key aspect for success are contributions from other users. Community members will have to register, but hide their professional identity from others, i.e., they can participate anonymously. By nature, all contributions are published and might help multiple visitors. For non-public support requests, some manufacturers are offering the option to submit a **private support ticket**. Each case will be handled one-to-one by a company employee and will be escalated to a product expert, if required.

Manufacturers of ASSP candidates in their field will be named later in corresponding chapters, but their sales partners typically work with different product lines from different manufacturers. Some of them are broadline catalogue resp. online distributors (marked with "O"), but others have dedicated and regional IoT sales and application support staff (marked with "S") and offer additional services, e.g., provisioning of secure MCUs. Table 4.1 provides a (non-exhaustive) list of distributors which are offering IoT ASSPs mentioned in this book (in alphabetical order).

Table 4.1 Distributors

Distributor Name	O = Online/Catalogue S = Service/Support	Homepage
Arrow	S	www.arrow.com
Avnet (incl. Abacus, EBV, Silica)	S	www.avnet.com
Digi-Key	O	www.digikey.com
Element14 (Farnell)	S	www.element14.com
Farnell	O	www.farnell.com
Future electronics	S	www.futureelectronics.com
Mouser	O	www.mouser.com
Newark (Farnell)	S	www.newark.com
RS Components	S	www.rs-components.com
Rutronik	S	www.rutronik.com
Sanshin	S	www.sanshin.co.jp
SOS Electronics	S	www.soselectronic.com
Symmetryst	S	www.symmetryelectronics.com
Wintech	S	www.wincomponents.com

4.3.2   IoT Clouds

By nature, IoT applications are connected to the Internet, exchanging data between IoT clients and an IoT server. During field operation, the IoT server will collect data from deployed IoT devices and perform further data processing according to application requirements. IoT data consolidation and analysis will generate actionable insights or remote-control commands to the IoT device, if applicable. In any case, the IoT server will have to be managed by the respective IoT business owner, but local operation might be difficult to cope for large-scale IoT deployments with many devices and big data load. Another challenge is to manage deployed devices in the field, i.e., to check status, update functionality or to re-configure them, if required. But also, during IoT device production, external services will be required to perform device provisioning and injection of an individual Root-of-Trust (see Sect. 4.2.4).

An IoT project can take advantage of external online IoT services instead of implementing them in-house, e.g., server hosting, a public-key-infrastructure (PKI), IoT device management, data analytics. An IoT cloud offers resources (servers, storage) and configurable functions for IoT applications and services to support deployment of IoT devices. In general, IoT cloud services are leveraging available external expertise of IT companies and offload inhouse development efforts to build an infrastructure for IoT device provisioning, management, and data processing. IoT cloud services allow IoT applications to select from a collection of options to collect, filter, transform, visualize, and act upon device data according to customer-defined rules.

Following increasing worldwide demand for IoT solutions, big IT players like Google or AWS have entered this market. In addition, some hardware manufacturers of cellular network modules or security MCUs are bundling IoT cloud and PKI services with their products in order to offer a one-stop-shopping experience to their customers. But in order to offer flexibility to their customers, most of them are also supporting 3rd-party IoT clouds (see Sect. 4.3.6 for details).

By nature, IoT clouds are versatile and generic, i.e., offered services are working independently from target application and device hardware or used network technology. Cloud services do interact on low-level directly with device hardware or operating system. Instead, on device side a secure network socket (typically TLS, see Fig. 4.4) is acting as a device identifier and endpoint for IoT cloud one-to-one communication. For most IoT clouds, HTTP or MQTT application layer protocols can be used for uplink IoT data transfer or for downlink device updates. Embedded HTTP and/or MQTT clients are standard firmware functions offered by most cellular network modules. This means that the device MCU will communicate with the device network socket via its AT command interface and manage all data transmission of the IoT device with the IoT cloud. Figure 4.11 is illustrating the interface and major IoT cloud functions.

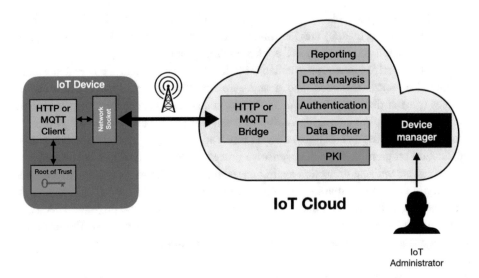

Fig. 4.11 IoT cloud interface

During initialization or field operation, a device management function can be used to set up or change individual device configurations incl. its Root-of-Trust, firmware updates, etc. It also maintains a logical configuration of each device and can be used to remotely control the device from the cloud. Since the IoT cloud does not know any technical details of the IoT device, each configuration request is must be converted locally by the host MCU into a sequence of module-specific AT commands. Available functions will be mainly determined by the device, e.g., by embedded software stacks and initial device provisioning. Third-party IoT clouds are offered by.

- Google Cloud IoT Core, URL: https://cloud.google.com/iot-core
- AWS IoT Core, URL: https://aws.amazon.com/iot-core/
- Microsoft Azure IoT, URL: https://azure.microsoft.com/en-us/overview/iot/
- IBM Watson IoT, URL: https://www.ibm.com/cloud/watson-iot-platform
- Telekom Cloud of Things, URL: https://iot.telekom.com/en/solutions/platform

and many others.

4.3.3 Host MCU

The host MCU has a central role because here the **embedded part of the IoT application** is located which determines the function of the IoT device and masters all external interactions. In particular, the host MCU controls the operation of the network interface ASSP (network sockets, protocols, etc.) via AT command interface. In addition, IoT peripherals (esp. sensors and actuators) are controlled, scheduled and payload data will be retrieved and pre-processed according to IoT application requirements.

For battery-driven operation, a suitable MCU should have efficient power-management functions, **low-power** modes, and a **wake-up function** to be triggered from external sources (pin). If dedicated functions for intrusion detection (see Sect. 4.1.3) are required, the MCU should offer an **analog-to-digital-converter** (ADC) and a **watchdog timer**. For mission-critical alert IoT devices, a **short interrupt latency** might be required. The IoT application program is containing sensitive processing information and intellectual property which is requiring protection against eavesdropping (refer to Sect. 2.4). An embedded **read/write protectable memory** which is large enough to store the IoT application program as well as IoT payload data and an event log file. A one-time-programmable (OTP) part of the internal memory is as plus for Root-of-Trust data or product lifecycle status. In addition, some MCUs are offering a solution for read-out protection of stored data, e.g. [5].

A simple low-cost general-purpose 8-bit MCU will have sufficient processing power to meet these functional requirements. Even public-key cryptography seems to be viable on 8-bit MCUs without hardware acceleration. According to a research paper [6], an Atmel ATmega128 at 8 MHz executes a 160-bit ECC point multiplication in 0.81 s and only needs 0.43 s for a RSA-1024 public-key operation—used for data encryption resp. verification of a digital signature. But for a data encryption incl. signing operation, duration will go beyond 10 s. And mentioned key sizes do not meet advanced security requirements resp. recommendations by federal agencies like NIST of 2048 bits for RSA resp. 256 bits for ECC (see [7]). Consequently, an **8-bit host MCUs (without a dedicated crypto accelerator) can be used only for low-security IoT devices or for IoT applications not requiring low latency**. On top of that, general-purpose MCUs have not been designed to resist against skilled, sophisticated attacks at level 3 or beyond (refer to Sect. 2.7). For advanced IoT security, the host MCU should have better crypto performance and offer improved protection. Figure 4.10 indicates two possible approaches:

- add a security coprocessor (secure element) or
- use a more powerful and security MCU.

4.3.4 Smart Card Security for IoT

The good news is that—for advanced IoT security—suitable and proven technology is available, but for different use case scenarios. In fact, **smart cards** are used for payment or personal identification purposes—as personal security tokens requiring ultimate tamper-protection in high-volume markets with a total annual demand of 10 billion units.

They are used as e-passports, health cards, e-tickets, SIM cards, etc. According to market research company GMI [8]), the global smart card market exceeded USD 40 billion in 2021 and still forecasted to grow 10% annually. This business has reached a high level of maturity, driven by global acceptance and security standards. Supply is dominated by a few semiconductor manufacturers with a long track record how to design secure microcontrollers to resist intrusion attempts by top-motivated attackers with highest level of expertise and budget. These companies are (in alphabetical order): Atmel/Microchip, Infineon Technologies, NXP Semiconductors, Samsung, and STMicroelectronics. Top smart card manufacturers are (in alphabetical order): Gemalto, Giesecke & Devrient, Morpho (Safran), Oberthur Technologies.

Smart cards are tiny electronic systems containing only one single component: a security MCU with a dedicated embedded software program. The semiconductor manufacturer will package each MCU die into 6- or 8-pin micromodules, electrically pre-configure and ship them to the smart card manufacturer for mechanical and electrical end-customer production and delivery (Fig. 4.12).

In fact, the smart card manufacturer is embedding the MCU micromodule into a conveniently sized piece of plastic which is also used for personalization, e.g., the card owner is printed here. But smart card manufacturers are also performing an electrical personalization of each card which is finally becoming a **unique personal token** which is being shipped to the card owner. Smart card manufactures are specialized to perform this kind of job for huge quantities. In order to meet advanced security requirements of governments, banks, etc. involved security IC production environment as well as smart card production environment will have to be secure and certified on behalf of business owners.

For IoT devices, personalization requirements are different, because identities reside in a piece of hardware which is mounted inside each device, instead of a removable card. Consequently, for IoT security ICs, many semiconductor manufacturers are taking over this part. In fact, each die (on a wafer) has to be qualified and tested electrically anyway, i.e., injecting a chip-individual identity to each die is just a minor extension of a standard test procedure for a semiconductor IC. And as a matter of fact, this production step

Fig. 4.12 Smart card

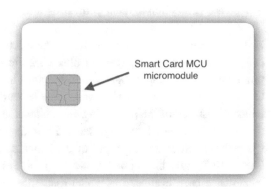

Smart Card MCU
micromodule

will be performed in a secure manner, if existing smart card IC production facilities and processes are being used also for IoT security chips. This way, IoT devices can benefit if **discrete hardware-based** solutions are being used to address IoT security requirements. This translates into cost savings as IoT device manufacturers do not need to invest in a dedicated infrastructure incl. their own secured production facilities and specialist know-how to deliver the highest levels of standards-compliant security to their customers.

But independent from these production aspects there is no doubt that hardware-based security solutions clearly outperform software-only approaches, e.g., where to store a Root-of-Trust for an IoT device. By nature, this is vulnerable data which is requiring strong protection against any access (e.g., for private keys) or at least against unauthorized write access. This is asking for a **secure memory** and a **secure processing environment** where credentials are being processed resp. where the encryption/decryption work is done. Potential data leakages should be eliminated in order to complicate elaboration of related attacks (refer to Sect. 2.4). **Certified security ICs** are offering state-of-the art protection measures and tamper-resistance—with the added acknowledgement of independent evaluations (see Secct. 2.10). In particular, a CC certification adds evidence that a security MCU is prepared to resist "type 4″ attacks (refer to Sect. 2.7) where sophisticated side-channel resp. fault-injection methods or invasive intrusion on chip-level are applied.

4.3.5   Secure Elements

Secure Elements (SE) are **discrete security modules** based on smart card IC technology. For an IoT device host MCU, a secure element works as a coprocessor with a dedicated high-speed crypto engine and secure data storage. In conjunction with the host MCU, the SE works as **Root-of-Trust of an IoT device**. In particular, it can be used to

- establish a **trusted communication channel** to external parties (device authentication, data encryption), esp. for interaction with the IoT server resp. IoT cloud,
- provide **secure storage** for sensitive data,
- increase **protection against pentesting** attempts (see Sect. 2.4).

A Root-of-Trust contains credentials for the electronic identity of each IoT device. Typically, this a unique public-key pair plus a certificate issued by a trusted third-party (CA) providing evidence that a signed message must have been originated by a known device. On top of this, the foundation of a reliable hardware-based Root-of-Trust is ultimate tamper-resistance of the processing environment and stored assets. Looking at some candidates at the time of writing (see Table 4.2), most SE manufacturers are offering a CC security certificate as a proof of implemented quality. But, as mentioned before (see Sect. 2.10.2), only the applied protection profile (PP) will tell, which functional requirements have been evaluated at which assurance level for each of them. For example, a

Table 4.2 Secure elements

Manufacturer	Product	CC Certification	URL
NXP	A5000	EAL6+	https://www.nxp.com/products/security-and-authentication/authentication:MC_71548
	SE050/SE051	EAL6+, SESIP4	
STMicroelectronics	STSAFE-A110	EAL5+	https://www.st.com/en/secure-mcus/stsafe-j100.html
Microchip	ATECC608B		https://www.microchip.com/en-us/product/ATECC608B
Infineon	Optiga Trust M	EAL6+(HW)	https://www.infineon.com/cms/en/product/security-smart-card-solutions

successful evaluation of security requirement AVA_VAN.5 (Advanced Methodical vulnerability analysis and resistance against high attack potential) provides extra confidence because the CC test lab has used a maximum level of expertise and state-of-the-art equipment to attack chip-internal assets. Nevertheless, all listed manufacturers have a proven smart card IC track record and corresponding design expertise, chip technology and secure production environments.

In addition, setting up a SE as a Root-of-Trust requires on-site provisioning and injection of a minimum set of credentials—for each chip. For this purpose, SE manufacturers can create **leaf certificates** for each SE chip, e.g., STMicroelectronics production CA signing the unique serial number of each STSAFE-A110. But a manufacturer CA can also qualify as an Issuing CA (see Sect. 4.2.4) to be part of the applied CA hierarchy (see Fig. 4.9) and associated chain-of-trust. For integration into existing cloud-based PKIs, most SEs offer built-in support for onboarding to popular **IoT clouds**. On top of that, most SE vendors are offering customer-specific on-site SE provisioning with custom credentials, if a certain minimum order quantity of some thousand units is reached. Otherwise, suitable PKI certificates will have to be self-created or purchased later, as needed. In any case, the chip-individual ID and certified PKI key pair will be a good starting point for integration into any given PKI structure, if needed.

Typically, host MCU and secure elements are using an **authenticated and encrypted I²C connection** for data exchange. Consequently, this tamper-proof data link works as a trusted MCU extension. Establishing this secure channel is called "pairing", and it is based on symmetric cryptography. For this purpose, two shared secret keys are used for mutual message authentication, one for host commands, the other one for SE responses. At the same time, each key will be used for data encryption to avoid eavesdropping of the I²C bus. In order to set up host pairing, both keys will have to be generated and securely loaded into the SE chip in a trusted manner. Typically, this initialization will be done during IoT device production or during device deployment. Later, during field operation and for extra security, the shared key can be exchanged by a new one at

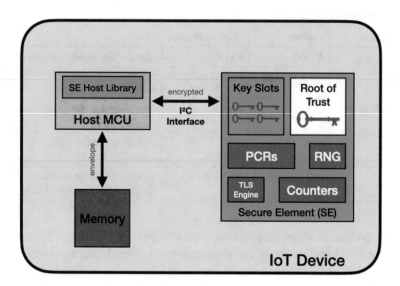

Fig. 4.13 Paired host MCU—secure element

any time by policy, e.g., once every day. Key generation is supported by physical **true random number generators (TRNG)** with certified quality according to NIST-800-90B. Figure 4.13 is illustrating a paired host MCU with a SE, and major functional blocks.

For secure external data storage, a **local envelope wrap/unwrap** mechanism can be used. The envelope is containing a secret key or a protected plain text, which has been encrypted by a symmetric wrap algorithm. For each envelope, a temporary symmetric key is generated and stored inside the SE. Wrapping/unwrapping commands are issues by the host MCU (after pairing with the SE) via secure I²C channel—and will be executed exclusively inside the SE in a secure environment. The envelope can then be used by the host MCU as needed, for example it can be stored or transmitted into unprotected environments. This feature, for example, can be used for event log files (refer to Sect. 2.5).

The SE command interface is provided by the SE manufacturer as a dedicated host library to be used by the IoT application, as needed. It will be provided as source code (e.g., for Infineon Trust M) or as an executable library (e.g., STSAFE-A110 host library for ARM-based MCUs). By nature, a secure element is a specialized MCU with dedicated accelerators for various cryptographic algorithms. Consequently, and in conjunction with a host library and a corresponding command set, the SE is a universal **cryptographic toolbox**. Functions include data encryption and decryption with symmetric and asymmetric keys for various algorithms and key lengths (RSA up to 2048 bit, ECC up to 512 bit, AES up to 256 bit, SHA hash functions up to 512 bit), different key agreement schemes (ECDSA, ECDH, ECDHE), random number and key generation, key derivation (HKDF), etc. (refer to the OPTIGA Trust M command set [9] as an example). Based on these capabilities, all SEs are offering support for the widely used application layer **TLS protocol** (see Sect. 3.6.1) and **HMAC scheme** (see Sect. 3.4.

In addition, an SE typically is offering some **monotonic counters** which can be used to limit user access to certain resources or an application. In general, an SE is offering a couple of low-level capabilities allowing designers to build a **custom crypto subsystem** according to project-specific requirements and/or a given PKI. For example, a SE chip offers a few kB of secure storage which can be used for keys and other Root-of-Trust credentials. So, for **IoT device platform integrity checks and secure boot**, SE-internal memory can also be used to store hashed device configuration data. Refer to Sect. 4.2.3. Or it, if integrity checks not needed frequently, PCR contents can be stored as a local envelope in unprotected host memory. A versatile concept how to implement SE-assisted device integrity verification is provided later in this book, see Sect. 5.2.3.3.

The Java-based NXP SE050/051 is a high-end SE and offers ultimate security which is CC-certified at EAL6+ incl. the highest level of vulnerability protection (AVA_VAN.5) against hardware attacks. The SE050 is also certified according to the SESIP evaluation scheme at level 4 with a hardware-based PSA-compliant Root-of-Trust. Due to its Java-approach, the SE050/51 is prepared to run application-specific embedded software but supposed to work in conjunction with a discrete host MCU like any other secure element (see Sect. 4.3.6).

For an IoT device with a high level of protection, a discrete, certified, and pre-provisioned secure element (SE) can significantly reduce project complexity and time-to-solution because all critical security design and production aspects of the IoT system design are covered by the supplied SE chip itself. Related cost savings might compensate additional spending of approx. 1 USD for a SE to be added to the bill-of-material for each IoT device.

4.3.6 Security MCUs

SE manufacturers can leverage existing security expertise originated from smart card markets but suffer from a lack of ASSP experience. ASSPs are addressing many customers, but smart card ICs are sold to very few smart card manufacturers only. This is making a difference because ASSPs require self-explanatory and comprehensive documentation, evaluation tools, application software and online support services (refer to Sect. 4.3.1). Vendors of general-purpose standard MCU are used to address these kinds of "mass markets".

In order to improve MCU security, owners of general-purpose MCU designs will have to add SE-like features to a standard MCU with no built-in security features, Fig. 4.14 is illustrating this idea. As a consequence, the embedded IoT application program will reside and will be processed inside a security MCU instead of a standard MCU. At first sight, this looks like a logical approach because now IoT security is located where the embedded IoT application needs it. And it will reduce bill-of-material and save cost.

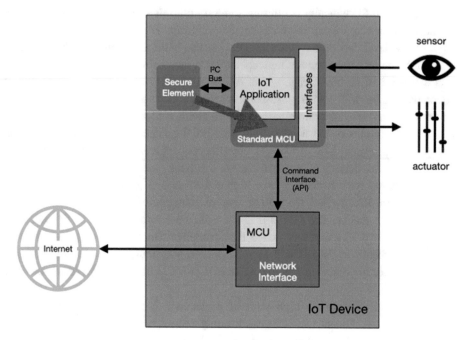

Fig. 4.14 Security MCU = standard MCU + secure element

But does this approach offer the same level of protection for IoT assets as a secure element? In case of doubt, a security evaluation by an independent party will provide evidence. In fact, most manufacturers of security MCUs try to avoid expensive and time-consuming Common Criteria evaluations, if a CC certificate is not mandatory for a specific project they would like to address, e.g., a mission-critical deployment of smart meters for energy consumption in households which is typically supervised by the respective national government. As an alternative or additional option for security MCUs, two **security evaluation** schemes have been established: SESIP and PSA Certified (see Sect. 2.10 for more information).

While the SESIP scheme is fully technology independent, PSA Certified evaluations are applicable to MCUs which are compliant to the Arm PSA Security Model [10]. PSA (Platform Security Architecture) defines a common hardware and software security platform, providing a generic security foundation and allowing secure products and features to be developed on top. First of all, PSA compliance means functional compliance which is dealing with interfaces, functional behavior, and interoperability—both for general product features as well as for security features. Although PSA Certified is platform-agnostic, it is no surprise that all PSA Certified-labeled MCUs (see complete list here: [11]) have been designed and manufactured by licensees of ARM processor cores.

For **PSA Certified security MCUs** three levels of evaluation are available. Both levels two and three evaluates the implemented PSA Root of Trust (PSA-RoT) in a test laboratory. **Level two** is aiming at IoT devices requiring resistance against software attacks only. Corresponding assets, threats and security functions are described in a 23-page document [12]. **Level three** is aiming at IoT devices requiring resistance against hardware and software attacks. Corresponding assets, threats and security functions are slightly extended by physical and side-channel attack scenarios and described in another 23-page document [13]. As an alternative, a SESIP Protection Profile can be used as a reference for evaluation. This applies, for example, to the STM32U585. Typically, certification at this level applies to a combination of silicon and firmware, see Table 4.3. This means that the certificate is valid only if both parts are used for a target IoT device. For this purpose, manufacturers have to provide guidance how to set up the secure solution properly, see [14] as an example for STM32U585 with TFM software.

Reference documents for SESIP certification of a security MCU are documents "Security Evaluation Standard for IoT Platforms" and "SESIP Profile for Secure MCUs and MPUs" (see [15]). Similar to CC, the SESIP profile determines security requirements and features leading to a Security Target (ST) which is describing the actual ToE (Target of Evaluation). Strength of implemented protection measures is determined at five assurance levels (see Sect. 2.10.3). Assurance levels 4 and 5 complement an existing CC evaluation including requirement AVA_VAN.4 (for SESIP 4) or AVA_VAN.5 (for SESIP 5) confirming that the ToE has already successfully passed a vulnerability analysis offers resistance against attacks with moderate (for SESIP 4) or high (for SESIP 5) attack potential. This option, at the time of writing, has only used for the SE50 product of NXP at SESIP Level 4, all other MCUs have been certified at SESIP Level 2 or 3.

NXP is following a special approach with their SE50/51 secure element. In fact, the SE50/51 has been rewarded by an EAL6+ CC-certificate with strongest vulnerability resistance of AVA_VAN.5 up to the embedded operating system level. But due to its Java-based operating system (JCOP) it offers the foundation for extra flexibility to develop embedded applications (applets). In fact, the SE051 allows to update the SE051's IoT applet with various items over-the-air, including applet updates or deploy newly developed applets after deployment. This means that, from a technical point of view, the customer IoT application could run on the secure element itself—removing the need for an extra host MCU. Unfortunately, at the time of writing, development of a new SE050/051 applet requires cooperation with NXP, i.e., the is no open applet development environment available e.g., for distribution customers.

SESIP and PSACertified security requirements are similar and based on the idea of an **immutable Root-of-Trust** which is containing the device ID, a unique cryptographic key, a boot ROM or other fundamental mechanisms that are not updateable. On top of this, a couple of security features are considered for evaluation:

- **Software Isolation**. A hardware-based mechanism that restricts access to a protected program (e.g., IoT application) by other embedded software.
- **Secure Storage** to protect integrity and confidentiality of sensitive assets
- **Security Lifecycle** to differentiate security properties for different platform states like vendor provisioning, device deployment, normal usage.
- **Attestation of Lifecycle State**.
- **Attestation of Genuineness** of platform or application software.
- **Cryptographic Functions** incl. generation of random number and keys.
- **Secure Boot**. A process to verify integrity and authenticity of executable code in a chain of trust starting from the Boot ROM.
- **Secure debugging.** Lock/Unlock debug ports.
- **Physical Attack Resistance**, either prevent or detect attacks incl. side-channel attacks requiring physical access to the ToE.
- **Secure update**, of platform or application software.
- **Rollback Prevention** prevents attackers to activate a previous firmware version with a known vulnerability.
- **Audit log** of internal cybersecurity events.
- **Information Purging** manages removal of specified data in case an attack has been identified internally.
- **Secure Communication** support for embedded application software.

Security MCUs are offering a robust silicon-based security foundation and tamper-resistance like active shielding or scrambled memory cells in order to complicate hardware attacks incl. microprobing. Other hardware security features like an immutable Boot ROM are available or code partitioning features in an effort to support protection of embedded software in a trusted execution environment (TEE). These measures are aiming at prevention of **software attacks**. In case of PSA-based MCUs, the hardware foundation for proper **isolation of protected firmware** is based on Arm's so-called "Trustzone" technology, see [16]. This TrustZone technology for Arm processors is the foundation for establishing a device Root-of-Trust based on PSA guidelines. It reduces the potential for attack by isolating the critical security firmware, assets and private information from the rest of the application.

Utilizing these features requires extra software layers and APIs allowing an efficient and hardware-independent development of IoT application software on a Root-of-Trust solution (see Fig. 4.15). Firmware-based API functions sometimes are called "middleware" and provide hardware-abstracted access to hardware-based security functions for cryptography, debugging/test interfaces, random number generation, secure storage. This kind of security core functions are needed for many application-layer features like remote device management, remote provisioning, device authentication, secure communication (e.g., TLS support), product lifecycle management, etc.

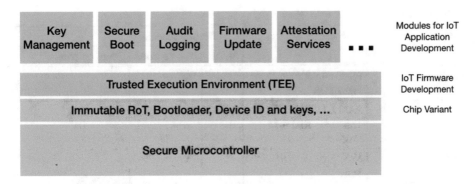

Fig. 4.15   Application interface layers

Table 4.3 shows a non-exhaustive list of certified security MCUs according to CC, SESIP and PSACertified evaluation schemes at the time of writing. Manufacturer websites are providing technical information as well as software support, evaluation tools and sales partners (distributors):

- Infineon: https://www.infineon.com/cms/en/product/microcontroller/32-bit-psoc-arm-cortex-microcontroller/psoc-6-32-bit-arm-cortex-m4-mcu/psoc-64/
- Microchip: https://www.microchip.com/en-us/products/microcontrollers-and-microp rocessors/32-bit-mcus/sam-32-bit-mcus/sam-l/sam-l10-l11
- Nuvoton: https://www.nuvoton.com/products/microcontrollers/arm-cortex-m23-mcus/ m2351-series/
- NXP: https://www.nxp.com/products/security-and-authentication/authentication:MC_ 71548 and https://www.nxp.com/products/processors-and-microcontrollers/arm-microc ontrollers/general-purpose-mcus/lpc5500-cortex-m33/lpc551x-s1x-baseline-arm-cor tex-m33-based-microcontroller-family:LPC551X-S1X
- Renesas: https://www.renesas.com/us/en/products/microcontrollers-microprocessors/ ra-cortex-m-mcus/ra6m4-200mhz-arm-cortex-m33-trustzone-high-integration-ethernet-and-octaspi
- Silicon Labs: https://www.silabs.com/mcu/32-bit-microcontrollers/efm32pg23-series-2
- STMicroelectronics (ST): https://www.st.com/en/microcontrollers-microprocessors/ stm32-32-bit-arm-cortex-mcus.html.

When **comparing security MCUs with Secure Elements**, in general, security MCUs are offering more complex security functions as secure elements which are quite limited to fundamental low-level embedded functions for cryptographic operations and secure storage, but SEs usually offer a higher level of certified security. Most certified security MCUs have been designed from scratch or derived from existing in-house MCU technology. For example, the Infineon PSoC 64 (originated from acquired Cypress Semiconductor) is

Table 4.3 Certified security MCUs

Manufacturer	Product	Certification scheme		
		SESIP	PSA certified	Common criteria
Infineon	PSoC 64 with Amazon FreeRTOS		Two	
Microchip	SAM L11 with Trustronic Kinibi-M TEE	2	One	
Nuvoton	NuMicro® M2351 Series		Two	
NXP	SE050	4		EAL6+
	LPC55S1x	2	Two	
Renesas	RA4M4	1	Two	
Silicon labs	EFR32FG23B SE Firmware V2.1.6, Secure Vault	3	Three	
STMicroelectronics	STM32U585 TFM, Version 1.0.0	3	Three	
	XCUBE SBSFU on STM32L476RG version 2.2.0	3	Two	EAL5+ (HW) for optional STSAFE-A110 secure element
Unisoc	UWP5663		Two	
Vango	Taishan400		Two	

based on the PSoC 6 which has a dual-core architecture and is predestinated for establishing isolated processing environments: The **secure processing environment (SPE)** and the **non-secure processing environment (NSPE)** run on different CPU cores. The SPE is keeping the immutable Root-of-Trust data, bootloader, etc. and offers secure services to the NSPE. Typically, the IoT application will be loaded for execution in the NSPE, but in case of the Microchip SAM L11 with Trustonic's Kinibi TEE (Trusted Computing Environment), the developer can add secure firmware modules and data, together with additional keys. In this case, these secure modules are programmed as "execute only" protecting their IP, whilst allowing factories further down the manufacturing chain to flash additional software. In general, all firmware modules will have to be assigned by the developer either to the secure world (SPE) or to the non-secure world (NSPE). For example, interrupt handlers should be executed in the SPE only, otherwise attackers might be able to modify them in a way that tamper events will be ignored.

In fact, most PSA-MCUs also offer security features aiming at **physical attacks**, even if they are not certified at this level, i.e., candidates are certified at PSA Certified Level 2 instead of Level 3. For example, Microchip SAM L11 is offering **peripheral access**

control (PAC) allowing to lock/unlock peripheral registers within the device in order to prevent malicious reconfiguration. PAC reports all violations that could happen when accessing a peripheral: write protected access, illegal access, enable protected access, etc. These errors are reported through a unique interrupt flag for a peripheral. On top of that, TrustZone technology is able to differentiate between **secure and non-secure peripherals** and can restrict NPSE access to memory locations as well as to peripherals. When a peripheral is allocated to the SPE, only SPE accesses to its registers are granted. In addition, both Microchip SAM L11 and NXP LPC55S1x are also allowing to **dedicate I/O pins** for secure communication with external devices and isolate them from non-secure applications—by pin multiplexing hardware.

But independent from TrustZone-specific security mechanisms, security MCUs are offering general-purpose countermeasures against physical attacks or pentesting (see Sect. 2.4). As a starting point, all of them have an **independent watchdog timer** which can be used for periodic operational checks and return the IoT application to normal execution, e.g., in case of a successful fault injection. Some MCUs are providing an **integrated ADC** which can be used to detect fault injection attempts, e.g., by measuring the applied power supply voltage. In combination with an external sensor, the ADC can also be used for physical tamper detection, e.g., by monitoring the case-internal humidity or incidence of light through an unexpected opening of the case (see related sample design in Sect. 5.2.2). Resolution of an integrated ADC will determine at which granularity decisions can be taken by the host MCU (see Fig. 4.16) if suspicious behavior is being detected. For same purpose, an integrated **temperature sensor** can be used, if available. Both ST MCUs are monitoring the integrity of the clock input as well the chip supply voltage internally. In order to complicate micro-probing of internal memory, address and data lines can be scrambled with a user-definable key (Microchip SAM L11). But **active shielding** of the MCU circuit will be detecting an intrusion anyway and trigger a user-defined action.

On top of internal security events, **tamper pins** can be used for external events. **Internal or external events** are indicating that something might be wrong and can be configured to trigger an interrupt or take immediate action. For example, an **automatic fast erase** of user-defined sensitive memory data can be performed (Microchip SAM L11 and STM32U585) or a direct memory access (DMA) request. In other cases, a suspicious event might require further qualification by the MCU. For this purpose, an **RTC (Real-Time Clock)** can be used to **count** detected events or determine its **duration** or **timestamp** an event for an audit log and later investigation. Quite unique is a security feature offered by NXP LPC55S1x called "code watchdog" checking the **integrity of executed code** in an effort to identify potential side-channel attacks.

By nature, all MCUs offer high-quality **true random number generators (TRNG)**, some of them are even certified (NXP and ST). On top of that, Silicon Labs EFR32FG23B and NXP LPC55S1x are offering a **physically unclonable function (PUF)** which can be used for identities and keys. Figure 4.16 is providing an overview about selected physical

	ADC	tamper-detection pins	tamper memory erase	sensors	RTC	Secure GPIOs (pin mux)	register write protection	watchdog timer	PUF	TRNG
Infineon PSOC 64	12-bit	-	-	temperature	yes	-	-	yes	-	yes
Microchip SAM L11	12-bit	yes	yes	temperature	yes	yes	yes	yes	-	yes
Nuvoton NuMicro M2351	12-bit	yes	-	-	yes	-	-	yes	-	yes
NXP LPC55S1x	16-bit	-	-	temperature	yes	yes	-	yes	yes	NIST SP800, FIPS 140-1
Renesas RA4M4	12-bit	-	-	temperature	yes	-	yes	yes	-	yes
Silicon Labs EFR32FG23B	16-bit	-	-	temperature	yes	-	-	yes	yes	yes
ST STM32U585	14-bit	yes	yes	temperature, voltage, clock	yes	-	-	yes	-	NIST SP800
ST STM32L476RG	12-bit	yes	-	temperature, voltage, clock	yes	-	-	yes	-	yes

Fig. 4.16 Countermeasures against physical attacks

protection and hardware-supported features of our candidates. Listed information is public data extracted from datasheets, etc.

Typically, manufacturers of security ICs **do not publicly disclose in-depth information about implemented countermeasures** against physical threats, e.g., which hardware shields are in place, how to defend against side-channel attacks, etc. More technical details about each product can be found in the Security Target (ST) document for CC or SESIP evaluations or certification reports. These documents are public and can be downloaded from the official CC website [17]. After all, the security certificate in combination with the applied Protection Profile (PP) will provide evidence about **which** security objectives have been met by a product—without telling **how** they these goals have been achieved.

Surprisingly, only few smartcard IC manufacturers are leveraging existing and CC-certified platforms for their security MCUs. Instead, STMicroelectronics has created a 2-chip security MCU solution containing an STM32 Arm-Cortex-M4 general-purpose 32-bit MCU combined with an CC EAL5+ STSAFE-A110 secure element, but—surprisingly—this solution has been SESIP evaluated at Level 2 only.

In-house **key provisioning and IoT cloud support** are offered by most manufacturers of security MCUs—similar to services offered for secure elements. This is a major advantage for IoT device manufacturers looking for a way to avoid investment in security-related design and production aspects. Secure provisioning services are also offered by some distributors, e.g., by Arrow Electronics for Infineon, Microchip and NXP security chips [18].

4.3.7   Secure Network Modules

Interfacing to a cellular network is complex and requires advanced RF and analog design expertise. On top of this, application developers expect a certain level of abstraction from complex 3GPP standards resp. from low-level knowledge of network physical layer, device-network synchronization, random-access procedures, etc. Know- how on this level is useful but not required for IoT application development. Instead, for efficient work, higher-level functions (API) and efficient tools are needed. Cellular modem vendors have recognized an increasing IoT demand from different industry segments, so they started to leverage their modem expertise for their offer of **comprehensive and easy-to-use subsystems.** These cellular network modules are for IoT application developers requiring cellular connectivity for their project without spending too much time with underlying cellular network technology itself.

In fact, core of each cellular network module is a **modem** (modulator-demodulator), i.e. a data converter which is modulating a carrier wave to encode digital data for transmission. In our case, transmission medium is a wireless cellular NB-IoT network with carrier frequencies of up to 2 GHz and output transmit power of up to 23dBm resp. 200 mW. This mix of digital, analog and power requirements means extra challenge for integration within a single semiconductor product. Thus, cellular network modules are usually containing a mixed-signal modem chip plus extra power amplifier and some other discrete components altogether in a compact multi-chip SMD package (e.g. a 96- pin LGA with 16×26x2.4 mm). This method of bundling multiple integrated circuits (ICs) and passive components into a single package is called System-in-Package (SiP). A typical cellular network module contains (see block diagram Fig. 4.17):

- Modem incl. command/data interface to IoT application (UART or USB)
- RF interface, amplifiers, filters
- Clock generation and distribution
- Power Management
- SIM card interface
- Microcontroller, OS, firmware, memory
- Analog/Digital Converter
- Peripheral interfaces (GPIOs, I^2C, SPI, etc.).

In fact, these network interface modules are very popular electronic components and the functional heart of most cellular IoT devices. According to market researcher Counterpoint, **shipments of cellular SiP modules have reached 100 million units in Q2/2021** and will cross 1.2 billion units by 2030. Leading manufacturers are Quectel, Fibocom, MeiG, Foxconn, Thales, Telit and Sierra Wireless [19]. Unfortunately, typical IoT network interface modules do not offer a secure processing environment nor strong protection of IP or other security assets. On the other hand, integrated software stacks offering secure

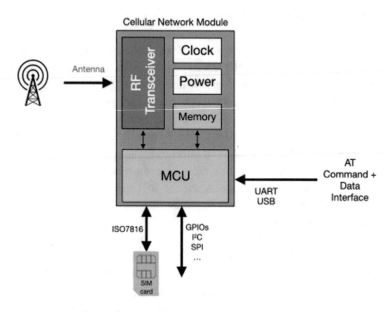

Fig. 4.17 Cellular network interface module—block diagram

communication protocols (see Sect. 3.6.1), secure OTA firmware function, etc. requiring cryptographic keys and random numbers. Consequently, for secure cellular IoT devices the network module will have to collaborate with an extra security MCU or secure element outside the network module.

For an IoT security point of view, an interesting alternative to a discrete solution based on a security MCUs or a secure element (SE) is offered by some manufacturers of network interface modules. The idea is to **integrate IoT security functions as well as the IoT application** into the IoT network module and reduce the bill-of-material of the IoT device, see Fig. 4.18. This way, only one MCU would be required for a secure cellular IoT device—instead of two or three MCUs. Result would be a **fully integrated secure IoT frontend** containing all IoT device core functions in a single component.

Due to the characteristics of its packaging, a multi-chip SiP module cannot offer the same level of protection against physical intrusion, microprobing incl. vulnerability analysis compared to a MCU subsystem integrating all functions of same silicon die like a security MCU or secure element. Consequently, at the time of writing, none of them has been certified according to a recognized security evaluation scheme (see Sect. 2.10). From a functional point of view, a blocking point for a fully integrated solution is that these modules typically have been built as a dedicated network interface—to be used by the IoT application via AT command interface. Typically, IoT network modules are not prepared to run user code and do not allow to develop a custom IoT application to be

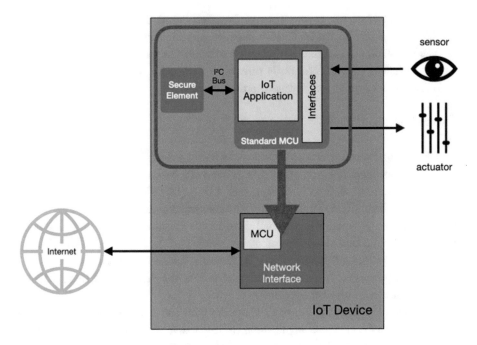

Fig. 4.18 Network interface module with embedded security

stored and executed by the module MCU. This option requires commercial development tools as well as dedicated manufacturer support.

Only few cellular IoT module vendors are meeting both criteria for a fully integrated and secure IoT network interface. This is a non-exhaustive list of cellular network interface modules allowing customers to **build an IoT device with a single central component**:

- **Nordic Semiconductor nRF9160**

  o URL: https://www.nordicsemi.com/Products/nRF9160
  o nRF9160 cellular IoT SiP offers LTE-M and NB-IoT connectivity as well as an integrated GPS/GNSS, a 12-bit ADC and various peripheral interfaces
  o Embedded Security: An Arm-Cortex MCU offers TrustZone technology for trusted code execution and an Arm Cryptocell as a Root-of-Trust and for cryptographic application-layer services.
  o The nRF9160 includes software for secure boot and secure over-the-air firmware update (FOTA) for application as well as for modem firmware.
  o The nRF Connect IoT software development kit allows development of custom application firmware. It integrates the Zephyr RTOS and a wide range of samples,

application protocols, protocol stacks, libraries and hardware drivers. nRF Connect SDK is publicly hosted on GitHub https://github.com/nrfconnect

o The nRF9160 offers a total of internal 1 MB flash memory

o For technical support, Nordic is offering a discussion board (forum), guides and online courses on https://devzone.nordicsemi.com

- **Sierra Wireless WP7702**

  o URL: https://source.sierrawireless.com/devices/wp-series/wp7702/

  o WP7702 cellular IoT SiP offers LTE-M and NB-IoT as well as 2G fallback connectivity integrated GNSS receiver, four ADCs (12/15-Bit) and various peripheral interfaces incl. I^2C, USB, GPIOs, wake-up interrupts, etc.

  o Includes professionally maintained open-source Legato platform (https://legato.io) for IoT application development of a Linux-based product. Code is available on Github https://github.com/legatoproject/legato-af#clone-from-github

  o WP7702 offers 512 MB flash memory with 256 MB dedicated to Legato platform incl. Linux and customer application

  o Embedded Security: Dual core CPU system with one Arm-Cortex A7 core entirely dedicated to the customer application. Built-in native security where all applications by default run in isolated sandboxes and have to be given permissions to interact with applications outside of the sandbox.

  o WP7702 incorporates Secure Boot and Secure Debug (disable RAM dump and JTAG port), FOTA via Sierra Wireless' AirVantage cloud service.

  o For technical support, Sierra Wireless is offering a discussion board (forum): https://forum.sierrawireless.com

- **Telit ME910C1 Series**

  o URL: https://www.telit.com/devices/me910c1-series/

  o ME910CE cellular IoT SiP offers LTE-M and NB-IoT as well as 2G fallback connectivity, optional GNSS receiver and various peripheral interfaces incl. I^2C, UART, USB, GPIOs

  o Telit AppZone software and application development. C programming language, Eclipse-based IDE, 4 MB file system, 1 MB App RAM

  o FOTA update function

  o Telit is offering one-to-one technical support via https://www.telit.com/contact-technical-support/.

Although the u-blox SARA-R5 cellular IoT network interface module does not offer custom application firmware development, it provides an extra level of IoT security:

- **u-blox SARA-R5**

  o URL: https://www.u-blox.com/en/product/sara-r5-series

  o SARA-R5 cellular IoT SiP offers LTE-M and NB-IoT connectivity, optional GNSS receiver (SARA-R510M8S) and various peripheral interfaces incl. I^2C, USB, GPIOs

  o <u>Embedded Security:</u> An immutable chip ID and hardware-based Root of Trust (RoT) embedded in a dedicated Common Criteria EAL5+ certified secure element provides foundational security and a unique device identity.

  > Secure libraries allow the (external) IoT application to use hardware-backed crypto functions and keys for local encryption to secure local file storage and for end-to-end encryption

  > Local data protection: symmetric crypto functions via AT command to locally encrypt /decrypt and authenticate data (e.g. certificates, tokens) on the device, also outside the module.

  > Cryptographic pairing between the IoT device host MCU and u-blox module by providing secure communication channel via UART interface.

  o <u>SARA-R5</u> offers secure boot and secure FOTA update

  o For technical support, u-blox is offering a discussion board (forum): https://portal. u-blox.com.

4.3.8 Physical Intrusion Sensors

Typically, IoT devices are the most vulnerable element of an IoT ecosystem and popular entry points for attacks (recall Sects. 2.3, 2.6 and 4.1.3). Attacker objectives depend on particular use case, but potential damage justifies investment to implement appropriate countermeasures. For example, a possible goal is to spoof stored credentials which are used by the device for authentication purposes. Based on this, an attacker can potentially clone a device with a criminal intention (e.g., unauthorized access to payable goods). Another goal is to modify transmitted consumption data in an effort to reduce corresponding bill. Or the intention is just to kill an IoT use case by taking out operational IoT devices (sabotage).

Physical accessibility of deployed IoT devices might allow local attacks, whereas availability of sample devices supports preparation of local or remote attacks (refer to Sect. 2.4). Both kinds of attacks are physical and usually start with attempts to open the housing of the IoT devices or get inside somehow in order to access internal electronic components or electrical interconnections. Mechanical countermeasures like one-way screws or sealing (see Sect. 4.1.1) might demotivate type-1 amateur attackers but will not stop criminals. Next level mechanical protection is utilizing **electrical switches or conductive mashes** attached to the internal surface of the product enclosure in a way that mechanical tampering will disconnect current flow through these detectors. Such an event can be used as an alert to the embedded IoT application, e.g., by submitting an interrupt to the host MCU. But pentesting analysis of a sample target IoT device will easily disclose how it works and how to work around it. More advanced intrusion protection is using **active sensing** of parameters inside the housing of the IoT electronics—by monitoring activity or suspicious conditions. In buildings, an infrared beam can be used which is triggering an alert when someone gets in the way, crossing the invisible barrier. Using a transmitter and a receiver, also ultrasonic time-of-flight techniques can detect distance or movement of persons, but for small "rooms" like the interior of an IoT device we need a tiny and inexpensive solution.

Active intrusion sensing for IoT devices is based on

- incident **light** or
- change of barometric **pressure** or
- detection of **noise** or
- detection of **vibration**.

For example, under normal operation conditions there will a stable pressure inside a closed housing, and it will be completely dark. Opening the case or drilling a hole will change internal pressure and brightness. Sensors are passive components; they can measure ambient parameters and report them. But sensors are not able to decide, if detected values actually indicate an intrusion or not. Instead, further qualification by the host MCU will be required, i.e., a **tamper detection algorithm**.

In fact, sensor outputs are either analog or digital signals. Common digital data interface is a serial I²C bus offered by most MCUs, for an analog sensor output signal an MCU with an integrated ADC is beneficial. Table 4.4 is listing major suppliers of sensors for most relevant intrusion indicators (ambient light, air pressure). All of them are available as a PCB-mountable version to be soldered into an IoT device:

Sensors for other parameters like vibration, temperature or acoustic signals are available and might be considered for expected attack scenarios and in combination with environmental conditions where devices are deployed. In general, selection depends on given IoT device design ingredients (e.g., host MCU), cost constraints and other requirements, e.g., if the device is battery-driven or not. In this case, **power consumption** of the sensor should be as low as possible, or it should offer a sleep mode allowing the sensor to weak up just in case the measured parameter exceeds a critical value. For this purpose, some digital sensors allow to set a certain threshold triggering the sensor to alert the MCU for further action. For analog sensors delivering a continuous stream of measurement data, an ADC will have to convert the output signal into digital format. The MCU will have to check other parameters for further evidence, that reported sensor data indicates an intrusion. Some MCUs (see Sect. 4.3.6) offer a versatile tamper-detection toolbox to support this task. In any case, the embedded IoT application will have to determine, if a detected

Table 4.4 IoT intrusion sensors

Manufacturer	Ambient light	Air pressure	URL
	A = analog out, D = digital out		
AMS	D		https://ams.com/en/ambient-light-sensors
Bosch Sensortec		D	https://www.bosch-sensortec.com/products/environmental-sensors/pressure-sensors/
Bourns		A+D	https://www.bourns.com/products/sensors/environmental-sensors
Broadcom (Avago)	D		https://www.broadcom.com/products/optical-sensors/ambient-light-photo-sensors
Honeywell		A+D	https://sps.honeywell.com/us/en/products/advanced-sensing-technologies
Infineon		A+D	https://www.infineon.com/cms/en/product/sensor/pressure-sensors/
Merit Sensor		A+D	https://meritsensor.com
NXP		A+D	https://www.nxp.com/products/sensors/pressure-sensors
Omron		A	https://components.omron.com
Panasonic		D	https://na.industrial.panasonic.com/products/sensors/pressure-force-sensors
ROHM Semiconductor	A+D		https://www.rohm.com/products/sensors-mems/ambient-light-sensor-ics
		D	https://www.rohm.com/products/sensors-mems/pressure-sensor-ics
Sensirion		A	https://sensirion.com/us/products/product-categories/differential-pressure/
ST Microelectronics		D	https://www.st.com/en/mems-and-sensors/pressure-sensors
TDK InvenSense		D	https://invensense.tdk.com/smartpressure/
TE Connectivity		A	https://www.te.com
Texas Instruments	D		https://www.ti.com/sensors/specialty-sensors/ambient-light-sensors/overview.html

(continued)

Table 4.4 (continued)

Manufacturer	Ambient light	Air pressure	URL
	A = analog out, D = digital out		
AMS	D		https://ams.com/en/ambient-light-sensors
Vishay	D		https://www.vishay.com/optical-sensors/digital-output-amb-sensor/
Würth Elektronik (WE)		D	https://www.we-online.com/catalog/en/wco/sensors

intrusion justifies immediate local action, e.g., to erase confidential data or other assets. This kind of "tamper-responsiveness" is critical and should be handled with extra care because it might turn the involved IoT device into a non-operational state. If preferred, operation might just continue, but a tamper event is logged internally or an alert message to an external party is submitted/transmitted. In any case, sensor **latency** might be another selection criterion, esp. if a sensor first has to wake up from low-power mode before any measurement process can start.

Section 4.3.8 is outlining a sample IoT device design with two sensors used in collaboration with tamper-detection features provided by the host MCU and coordinated by embedded software.

4.3.9 IoT Device Production Stages

On its way from production to field operation, an IoT device is passing several **maturity stages**. Security MCUs, secure elements and sensors are off-the-shelf "standard" chip products, which means that the same "off-the-shelf" product is addressing the needs of many customers. Customization is done by configuration of a standard chip according to application requirements. For typical standard products, this software adjustment is part of the system design and will be managed during production on customer side. But for tamper-proof devices, production requirements are more complex because the manufacturer has to generate secret data (e.g., device identities, cryptographic keys) and securely exchange data with external parties (e.g., with certification authorities). Usually, production facilities and processes of equipment manufacturers are not prepared to handle these security aspects. Consequently, most semiconductor companies are offering to handle sensitive ID and PKI data as a service called **provisioning**. In particular, this applies to manufacturers of smart card ICs with **CC certified production sites** providing secure processes and infrastructures, e.g., for payment or ID cards. Leveraging this capability, the IoT device manufacturer can outsource this sensitive part of the production process and buy provisioned ready-to-use MCUs. Now, these security parts can be handled by their production department like any other electronic component. See Fig. 4.19.

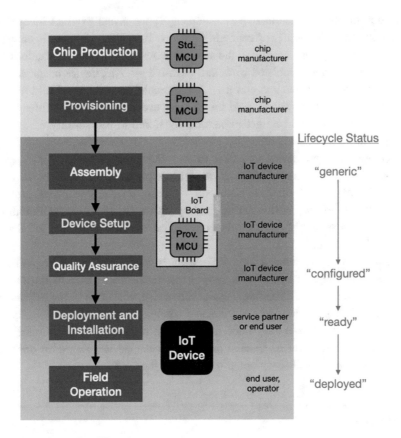

Fig. 4.19 Sample product lifecycle

After PCB board assembly on system manufacturer side, each new device has to be prepared for delivery, step by step. First, and in order to activate the full set of functions, a dedicated **device setup** program will have to configure all key components for proper cooperation. For example,

- pre-provisioned data will have to be relocated within the system or
- some components require "pairing" (e.g., MCU and SE, see "SE-centered concept") or
- calibration of sensors might be required or
- an initial integrity measurement has to be performed (see Sect. 4.2.3) or
- a country-specific product version has to be configured requiring different default parameters, e.g., for cellular network connectivity.

Finalization of this production step is incrementing the actual **product lifecycle stage**, for example, from "generic" to "configured" status. Each lifecycle stage is enforcing an dedicated security policy specifying the security status as well as authorizations for different user groups and associated **roles and permissions**, e.g., device manufacturer, IoT operator, service technician and end user. For example, an end user should not be able to perform a firmware update or read the security event log. The actual lifecycle status should along with related parameters be securely stored in permanently write-protected memory (e.g., in OTP).

A "configured" new device will have to pass functional and quality tests, as usual. Then, as a final production step, the device will be prepared for end user delivery. But before assigning "ready" lifecycle status, all test ports (JTAG, SWD) must be disabled, interfaces which are not needed during field operation should be closed and all tamper countermeasures should be enforced. Now, based on this final setup, the IoT device can be sold and deployed.

For some IoT applications, device deployment is critical and should be performed by authorized staff only. In this case, a service technician might have to finetune device location (e.g., for reliable operation), mount the device, connect it to the IoT server, etc. and make sure that it works properly. Related configuration parameters (e.g., connected network cell) will be fixed and stored along with final integrity measurement. Then, final "deployed" label can be granted to this particular device. Based on this lifecycle status, the IoT device will start normal operation in the field.

References

1. Ruhr Universität Bochum, PHYSEC. Enclosure-PUF. (2019). Retrieved March 29, 2022, from https://hardwear.io/netherlands-2019/presentation/Enclosure-PUF-hardwear-io-nl-2019.pdf.
2. Quectel. (2022). BG95&BG77&BG600L Series TCP/IP Application Note. Retrieved May 9, 2022, from https://www.quectel.com/wp-content/uploads/2021/04/Quectel_BG95BG77BG600L_Series_TCPIP_Application_Note_V1.1.pdf.
3. Khraisat, A., & Alazab, A. (2021). A critical review of intrusion detection systems in the internet of things: techniques, deployment strategy, validation strategy, attacks, public datasets and challenges. *Cybersecur, 4*, 18. https://doi.org/10.1186/s42400-021-00077-7.
4. Design And Reuse. (2022). Understanding Physical Unclonable Function (PUF). Retrieved May 13, 2022, from https://www.design-reuse.com/articles/47717/understanding-physical-unclonable-function-puf.html.
5. STMicroelectronics. (2022). AN4758 Application Note. Proprietary code read-out protection on STM32L4, STM32L4+, STM32G4 and STM32WB Series MCUs. Retrieved June 18, 2022, from https://www.st.com/resource/en/application_note/an4758-proprietary-code-readout-protection-on-stm32l4-stm32l4-stm32g4-and-stm32wb-series-mcus-stmicroelectronics.pdf.
6. Gura, N., Patel, A., Wander, A., Eberle, H., & Shantz, S. C. (2022). Sun Microsystems Laboratories. Comparing Elliptic Curve Cryptography and RSA on 8-bit CPUs. Retrieved May 18, 2022, from https://people.eecs.berkeley.edu/~pister/290Q/Papers/Security/ECC_small_dev.pdf.

7. NIST. (2022). *Recommendation for Key Management, Part 3: Application-Specific Key Management Guidance.* Retrieved May 18, 2022, from https://nvlpubs.nist.gov/nistpubs/specialpublications/nist.sp.800-57pt3r1.pdf.

8. GMI—Global Market Inside. (2022). *Smart Card Market.* Retrieved May 18, 2022, from https://www.gminsights.com/industry-analysis/smart-card-market.

9. Infineon_OPTIGA_Trust_M_Solution_Reference_Manual_v3.15. (2022). Retrieved May 25, 2022, from https://github.com/Infineon/optiga-trust-m/blob/develop/documents/OPTIGA_Trust_M_Solution_Reference_Manual_v3.15.pdf.

10. Arm PSA Security Model. (2022). Retrieved May 25, 2022, from https://www.psacertified.org/app/uploads/2021/12/JSADEN014_PSA_Certified_SM_V1.1_BET0.pdf.

11. PSA Certified Products. (2022). Retrieved May 25, 2022, from https://www.psacertified.org/certified-products/.

12. PSACertified. (2022). Lightweight Protection Profile—level two. Version 1.2. Retrieved March 26, 2022, from https://www.psacertified.org/app/uploads/2022/05/JSADEN002-PSA_Certified_Level_2_PP-1.2.pdf.

13. PSACertified. (2022). Lightweight Protection Profile—level three. Retrieved March 26, 2022, from https://www.psacertified.org/app/uploads/2020/12/JSADEN009-PSA_Certified_Level_3_LW_PP-1.0-BET02.pdf.

14. STMicroelectronics. (2022). STM32U585xx security guidance for PSA CertifiedTM Level 3 with SESIP Profile. User manual. Retrieved June 10, 2022, from https://www.st.com/content/ccc/resource/technical/document/user_manual/group2/ef/50/a5/51/b3/ab/4b/e3/DM00778539/files/DM00778539.pdf/jcr:content/translations/en.DM00778539.pdf.

15. SESIP Profile for Secure MCUs and MPUs. V0.0.0.7. (2021, June). Retrieved March 20, 2022, from https://globalplatform.org/wp-content/uploads/2021/06/GP_SESIP_Profile_Secure_MCU_MPU_v0.0.0.7_clean.pdf.

16. ARM Technologies. (2022). Trustzone for Cortex-M. Retrieved March 29, 2022, from https://www.arm.com/technologies/trustzone-for-cortex-m.

17. Common Criteria. (2022). Certified Products. Retrieved June 3, 2022, from https://www.commoncriteriaportal.org/products/.

18. Embedded. (2022). Arrow Electronics extends global secure provisioning services for IoT devices. Feb 2019. Retrieved June 18, 2022, from https://www.embedded.com/arrow-electronics-extends-global-secure-provisioning-services-for-iot-devices/.

19. Counterpoint. (2022). Global Cellular IoT Module Shipments to Cross 1.2 Bn Units by 2030. Retrieved June 3, 2022, from https://www.counterpointresearch.com/global-cellular-iot-module-forecast-2030/.

Device Design

5

The IoT success story is a global success story with contributions from different markets and for different use cases all over the world. International standards as well as global supply with standard products are helping to offer cost-efficient IoT solutions. Same applies to IoT security requirements and globally accepted security certification schemes like Common Criteria, SESIP and PSA Certified. But regional IoT markets are following different legal, cultural, and political frameworks leading to different business models and different technical product requirements. This might change over time, but for the time being, different regional preferences are blocking evolution of common IoT device platforms which can be used worldwide. Today, regional product and security requirements for IoT devices are different, eve1-6n if they address the same target application (e.g., for power consumption meters in households).

But some IoT security design elements are universal and unquestioned whenever they are able to **protect a commercial IoT deployment and reduce risk of financial loss** for their business owners. Many standards and design ingredients have been discussed in previous chapters, now we take a close look at practical design aspects for secure IoT devices with a focus on selection and use of off-the-shelf standard products.

5.1 MCU Options for Secure IoT Devices

First of all, choice of key components for a new IoT device design will follow functional requirements of the target IoT application. For example, selection of a suitable network technology will be driven by the IoT use case and deployment areas. Sensors will be determined by remote parameters to be captured. Further application-specific requirements include device power consumption, processing power, memory size, temperature range,

© The Author(s), under exclusive license to Springer Nature Switzerland AG 2022
K. Heins, *Trusted Cellular IoT Devices*, Synthesis Lectures on Engineering, Science, and Technology, https://doi.org/10.1007/978-3-031-19663-8_5

etc. Besides a limited engineering budget, also product cost (BOM) and manufacturing cost are further constraints.

But in general, IoT device designers have several MCU options, i.e., where to store and execute the embedded IoT application program and how to ensure reliable and trusted operation of the IoT device. **Integrated solutions** are offering comprehensive execution environments with less chips, but they are also less flexible than **discrete solutions**. following options are available:

- **Standard MCU** (general-purpose MCU, see Sect. 4.3.3)
- **Security MCU**
- Standard **MCU plus Secure Element (SE)**
- **Integrated Secure Network Module**.

As a starting point for less advanced security requirements, a standard MCU can be used, but a basic set of fundamental rules should be followed in order to minimize risk of damage without adding extra cost, see Sect. 5.2.1. Resulting protection level will be sufficient to resist at least type-1 attacks (see Sect. 2.7).

Highest integration for use in small form-factor IoT devices is offered by network interface modules allowing users to develop and run a custom IoT application on top of provided network functions (see Sect. 4.3.7). From an IoT security point of view, these integrated solutions offer a natural advantage vs. a discrete MCU because less entry points for pentesting or attacks are provided. Some of them are based on an Arm CPU core offering PSA incl. TrustZone capabilities which is qualifying them to offer a same level of IoT than security MCUs (see Sect. 4.3.6), but none of them is certified. In fact, a certified security MCU is offering excellent value for IoT devices complying with highest IoT security standards. Provisioned versions of security MCUs are offered to prepare IoT devices for use in a public-key-infrastructure (PKI) and with IoT clouds. But for ultimate type-4 protection and qualification for national rollouts, an IoT device will have to integrate a smart card module or a secure element (see Sect. 4.3.5).

5.2   Design Tips

The following chapters are outlining some concepts and suggestions for different design aspects of secure IoT designs with standard MCUs, a security MCU as well as a security concept a design which is based on a secure element (SE).

5.2.1 Fourteen (14) Tips How to Improve Security of Any IoT Device

From a functional point of view, many simple and inexpensive general-purpose MCU are able to run the embedded part of an IoT application, but they do not offer any built-in security features, secure storage and might not even have sufficient processing power for cryptographic operations. So, in any case, selection of an appropriate host MCU should follow security requirements based on a decent threat analysis and risk assessment (refer to Chap. 2). Selection of a suitable MCU is key for implementation of proper IoT device security. But nevertheless, keeping some fundamental security aspects in mind will help to put a minimum level of protection in place, probably good enough to resist type-1 attacks or to recover from an attack. Here is a list of suggestions for proper MCU configuration and functional software add-ons which are applicable to many off-the-shelf standard MCUs, easy to implement and free-of-charge. Refer to Sect. 2.5.

1. **Secure MCU configuration for deployment.** After finishing device design and manufacturing stages, all MCU configuration settings will have to be checked one-by-one in order to make sure that default settings do not cause vulnerabilities in the field. For example, MCU manufacturers are providing various tools for software development or for quality assurance allowing to access internal resources. So, the manufacturing process has to consider different product lifecycle stages and roles in order to exclude production tools and limit access options to IoT devices after shipment (see Sect. 4.2.4)

2. **Handling of device interfaces.** Attackers are always looking for entry doors how to access MCU-internal resources like memory location or registers. In order to limit vulnerability, all unused GPIOs and bi-directional data interfaces should be turned off or configured as outputs (see Fig. 5.1). Used interfaces (e.g., the UART command interface) should be disabled during inactive periods. Any input data should be checked carefully.

3. **Remove or disable not-required data functions.** MCUs and network interface modules are offering a couple of protocols and remote-control functions like Telnet, FTP which are allowing users to access internal resources. If not needed for the actual IoT application, these functions should be removed from firmware or disabled (see Fig. 5.1).

4. **Remote firmware update.** In case of a critical security breach (e.g., a tampered device), a new software version for the involved device(s) sometimes can fix the problem efficiently and resume operation. Also, bug fixes or functional improvements might be required occasionally. For deployed IoT devices, a local update or replacement would be too expensive and take too long. Remote device updateability is crucial. A roll-back mechanism must ensure that there is no way to re-activate a previous firmware version.

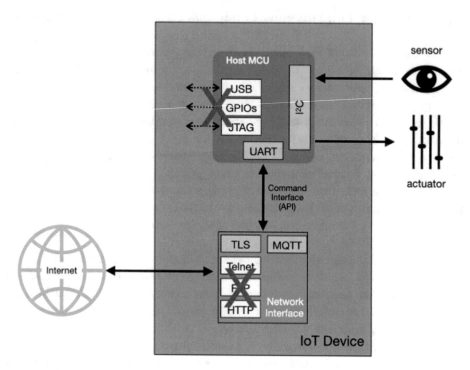

Fig. 5.1 Remove entry doors for attackers

5. **Verify firmware integrity** before execution. Only software from a trusted source is
   supposed to run of the IoT device. For this purpose, a valid firmware version has to
   be hashed and signed by the manufacturer, and an immutable (e.g., ROM-based)
   IoT device bootloader should verify this signature with a matching pre-installed
   immutable cryptographic key. If only simple key management is used, this process
   will quite efficiently reject execution of unauthorized software. Software tampering
   is still not impossible, but much more difficult to execute.
6. **Immutable Bootloader.** In order to start program execution from a trusted starting
   point, the initial boot code must be 100% reliable. Even if the device firmware should
   be updateable, this part (bootloader) must be fixed (resp. one-time-programmable) and
   stored securely inside each device during production.
7. **Device Identity.** Each IoT device is associated to a person or referring to a specific
   location where it has been deployed. The IoT operator must be able to identify each
   device remotely. This device identifier must be unique and immutably stored inside
   each device before it is getting deployed.
8. **Secure external communication channels.** In order to prevent eavesdropping and
   tamper attempts, all sensitive data leaving or entering the IoT device should be

unreadable for unauthorized external parties—using a simple and fast encryption scheme with a shared secret key.

9. **Use secure memory for sensitive data or IP.** By nature, on-chip MCU memory is less vulnerable against eavesdropping than external memory. Most sensitive data and software (e.g., keys and intellectual property) should be kept inside and should never be used externally and should be configured as read-protected, if possible. If larger secure memory is required, external data should be encrypted by the MCU before.

10. **Fault-tolerant programming.** Software countermeasures cannot substitute SE capabilities in this particular field, but code redundancy and error detection techniques can help to resist faults including attack-generated faults (see Sect. 2.4.2). Another good idea is to mitigate key spoofing damage by limiting lifetime of cryptographic keys resp. using session random keys instead, when possible.

11. **Use a watchdog timer.** Most MCUs are offering an independent watchdog which is a free-running down-counter. Once running, it cannot be stopped. It must be refreshed periodically before it causes a reset. This feature provides a good fallback for the IoT program whenever an unexpected event has caused a system failure—e.g., by an attacker. After restart, it might be possible to recover the device and reactivate it without having to terminate operation or to replace it.

12. **Use on-chip system resources and functions.** Besides memory and a real-time clock (RTC) other internal resources should be used, if available. For example, an internal clock source and internal voltage regulators are reliable and less vulnerable. Same applies to a built-in random-number-generator (RNG) which typically offering better entropy vs. software solutions.

13. **Sign of life.** For non-interactive and event-driven IoT devices, the operator must be able to request a simple sign-of-life indicator or status report from the device. This reply will provide evidence that the device is still operational and works as expected.

14. **Maintain a cybersecurity event log.** All tamper detections or other suspicious events or activities should be time-stamped, and all relevant parameters should be stored securely in a non-volatile memory inside the device. This log file can be used later as tamper evidence or for investigation purposes by the IoT operator.

5.2.2 Physical Intrusion Protection

As explained in Sect. 4.3.8, active monitoring of case-internal parameters adds an extra protection level on top of simple mechanical means like one-way screws or a sealed enclosure. The actual IoT use case in combination with actual locations where IoT devices are deployed will determine which kind of attacks are expected and implemented countermeasures will have to address these potential attacks accordingly. For the sample protection concept presented in this chapter, we assume that **incident light and change of barometric air pressure** inside the IoT device are most reliable indicators for a physical intrusion attempt. Another assumption is that the targeted IoT market does not require ultimate protection of stored assets nor resistance against side-channel attacks nor a CC or SESIP security certification on top level with a PP asking for a vulnerability resistance of AVA_VAN.4 or 5. In this case, a secure element must be used (see Sect. 4.3.5 or design concept "SE-centered concept"). On the other hand, implemented IoT security measures should be good enough to resist type-3 attacks (refer to Sect. 2.7).

These requirements, for example, are met by an STM32U585 to be used as the host MCU of the IoT device. The STM32U585 is member of the popular STM32 family of general-purpose Arm-Cortex MCUs with enhanced security features and certified according to PSA Certified Level 3 with SESIP Profile (see [1]). On top of standard PSA features (incl. TrustZone, etc.) aiming at prevention of software attacks, specific anti-tamper features are used to protect sensitive data from physical attacks, see Fig. 4.16. In case of tamper detection, sensitive data in parts of internal memory, caches, and cryptographic peripherals can be immediately erased, and a reboot is forced, if the STM32U585 is configured accordingly. Both external active tamper pins and internal tamper events are used. The **external tamper pins** can be configured for edge detection, or level detection with or without filtering, or active tamper that increases the security level by auto checking that the tamper pins are not externally opened or shorted. In latter case, the anti-tamper unit (TAMP) of the STM32U585 MCU will continuously compare output TAMP_OUT with input level of TAMP_IN (see Fig. 5.2) in order to detect removing a fastener or drilling a hole into the product enclosure which is protected by an embedded conductive mesh.

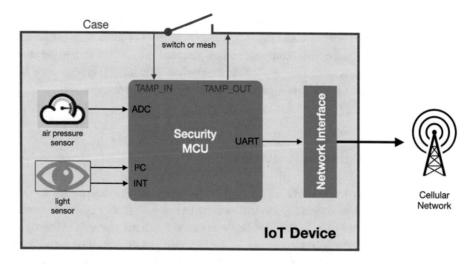

Fig. 5.2 Physical intrusion protection—block diagram

The integrated ADC of the STM32U585 can be used to monitor the internal air pressure. For this purpose, an **analog pressure sensor** of Omron, Sensirion or TE Connectivity can be used, see Table 4.4. For detection of incident light due to a physical breach of the device case, a **digital ambient light sensor** should be used, e.g., the Vishay VEML6030 [2]. The VEML6030 is digital 16-bit resolution sensor in a tiny 2 mm × 2 mm package. Besides an I^2C bus communication interface it offers a configurable interrupt feature. An interrupt is triggered whenever the actual measurement value crosses boundaries of a programmable 16-bit threshold window. This feature can be used to trigger an MCU interrupt handler for further investigation via I^2C data interface under software control, see Fig. 5.2.

But in general, careful analysis and qualification of input data will be required to classify them as a potential tamper event. For critical sensor data which might require immediate reaction, mentioned **anti-tamper unit (TAMP)** should be used instead of the internal ADC in combination with an interrupt trigger—because resulting tamper response time will be delayed by MCU interrupt latency time. Instead, for mentioned pressure sensor delivering measurement data via analog input, an ADC watchdog can be used to trigger the TAMP unit. In fact, this peripheral is part of the V_{BAT} domain and also contains the real-time clock (RTC), an internal 32 kHz clock source and battery-powered backup registers. The V_{BAT} pin allows the device to be powered from an external battery or an external super-capacitor, and V_{BAT} operation is automatically activated when V_{DD} is not present, e.g., in case the power supply has been deactivated by an attacker. Consequently, this V_{BAT} domain is always operational and allows either immediate erase of memory locations or double-checking of internal or external **tamper events** before the MCU is getting interrupted and software control is being triggered. Figure 5.3 is illustrating how tamper attempts can be handled by the STM32U585 MCU.

Several **internal sources** can cause an internal tamper event, mentioned ADC watchdog is one of them. The system clock as well as the power supply are being monitored in order detect an intentional voltage reduction or blocking of the MCU clock if an attacker tries to freeze the device state. Another example is an unauthorized use of a JTAG or SWD test port to access internal MCU resources. Of course, also an expired watchdog timer will also be handled as an internal tamper event. Typical external tamper events are managed by configurable tamper pins for use with different switch and mesh constellations and an anti-glitch filter.

Once a tamper has been detected, several different responses are possible. Each source of tamper in the device can be configured to trigger the following events within V_{BAT} domain without interfering with the regular operation of the IoT device:

- trigger a low-power timer
- erase keys and other device secrets in crypto units, backup registers or SRAM locations, MCU instruction and data cashes
- use the RTC to create a timestamp for the event
- alert external peripheral devices via GPIO output pin.

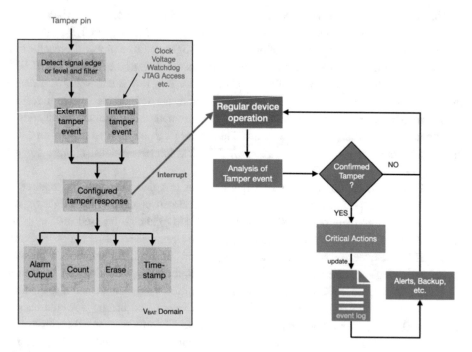

Fig. 5.3 Tamper handling overview

These actions are not delayed by any MCU interrupt latency and will allow quickest possible tamper responsiveness in case immediately action is required to prevent damage. But if further qualification of the potential tamper event is required, the MCU should take over and perform further analysis and cross-checks with additional data sources incl. other sensors. For this purpose, the TAMP unit can be configured to generate an **interrupt** which is also capable of waking up the MCU from a power-down or stop mode.

As a consequence, **regular operation of the IoT device will be suspended** and the interrupt handler will take over. Additional data and further insights should be considered to confirm the criticality of the potential tamper event. This analysis must lead to an appropriate set of tamper responses, e.g., erase or relocate sensitive data, trigger a built-in actuator, send an SMS alert to the IoT operator, transmit data to the IoT server. Depending on the nature of the attack, only a limited period of time will be available to act effectively. A reasonable **prioritization of tamper responses** might be required. In order to prevent damage, first priority is to enforce the technical countermeasure against the attack, e.g., to rescue system-relevant assets or erase intellectual property as well as device-individual credentials and secret keys allowing an attacker to torpedo the whole IoT application, clone devices, etc. Then, every potential tamper event should be timestamped by the RTC and kept in a log file which is stored in a safe, non-volatile memory location. Last but not least, stakeholder alerts should go out, and other less critical activities should be performed—hoping that the IoT device is still operational and will be available for execution, e.g., to send a message or transmit backup data via the cellular network.

5.2.3   SE-Centered Concept

Security MCUs (see Sect. 4.3.6) are offering a comprehensive security package which is addressing all major IoT security aspects, so they are certainly a good choice for most IoT applications, esp. for companies entering this market with new business ideas. Same applies to security-enhanced all-in-one network interface modules allowing users to use the integrated MCU for their own custom IoT application code (see Sect. 4.3.7). But in most cases, existing IoT embedded software originated from previous projects will have to be ported to a new CPU core, typically to an Arm Cortex CPU with TrustZone architecture which is used for most security MCUs. But many IoT device design projects probably will not suffer from this potential disadvantage because Arm is the leading IP provider for CPU cores, and ARM licensees dominate the MCU market for embedded systems, i.e., using an Arm Cortex CPU is no blocking point. But for cost-sensitive IoT devices, price surcharge for security-enhanced TrustZone Arm MCUs vs. standard 32-bit Arm MCUs might be a problem.

For other target applications, esp. governmental controlled large-scale rollouts like for electricity smart meters, lack of an **acknowledged security certification** are excluding most security MCUs from qualification. These projects typically require ultimate protection against up to type-4 attacks (refer to Sect. 2.7) and an augmented CC EAL4+ certification including extra resistance against attacks with highest potential like microprobing and SCA/FI attacks (see Sect. 4.1). Secure MCU do not meet these advanced security requirements, even a PSA Certified Level 3 certificate has been achieved. Instead, state-of-the-art protection of embedded software and data assets is offered by smart cards or secure elements only.

For ultimate IoT security and maximum flexibility, the following concept is outlining how to **combine any MCU with a secure element** in an effort to establish a Root-of-Trust (RoT) for an IoT device. For this project, an Infineon OPTIGA Trust M secure element with CC EAL6+ certified hardware has been selected, refer to [3] for the complete set of product documentation and resources. An OPTIGA Trust M is available for less than one USD per unit as a pre-provisioned SE with trust anchors and key certificates for PKI integration and IoT cloud onboarding (e.g., for MS Azure or AWS) of the IoT device. During production at Infineon fab,unique asymmetric keys (private and public) are generated and symmetric key/shared secrets are provisioned. The public key is signed by a CA and the resulting X.509 certificate is securely stored inside the SE. Custom provisioning is offered by Infineon and via distribution partners, e.g., Arrow Electronics.

On top of this, the SE offers a versatile high-performance crypto toolkit for use by the customer IoT application, e.g., for generation of random numbers and keys, digital signature operations, data encryption, etc. In addition, the SE provides a small secure internal memory of 10 kB (4.5 kB are available for user data) but allows to establish a large shared external memory with encrypted contents, e.g., for an event log.

This two-chip solution offers a lot of flexibility and—in combination with a PKI—an IoT application with strong protection against various logical attacks can be built. Main purpose is to provide strong mutual authentication, secure data storage and transmission incl. firmware updates, but it can also be used to ensure **system integrity and secure boot**. For Fig. 5.5 illustrates the host-SE interface and major functional blocks.

The SE chip connects to the host MCU via **I^{2}C interface** which is allowing to interact securely with the SE. Interaction is based on **SE host library** which provided as a free open-source MIT license which is a short and simple permissive license with conditions only requiring preservation of copyright and license notices, no other restrictions apply. C source code incl. a reference implementation of the OPTIGA Trust M host library (for Infineon XMC4800) is available in a GitHub repository (see [4]) and can be converted/ported for use on any target host MCU.

5.2.3.1   Secure I^{2}C Connection

The process of **pairing the MCU with the SE** allows to put a mutually authenticated encrypted command interface in place. Pairing requires presence of a **pre-shared secret** key ("Platform Binding") on both sides. This key will be used to generate a temporary symmetric crypto key (AES 128) for each data encryption/decryption session. Consequently, agreement of the pre-shared secret has to be done only once—during IoT device production in a trusted environment (see Sect. 4.3.9). The SE can generate this pre-shared key (random number) and export it (via I^{2}C interface) for storage in non-volatile memory of the host MCU, i.e., in a one-time-programmable (OTP) or write-protected area. Later, during device operation, the pre-shared secret is processed internally and will never leave the chip, but for extra tamper-resistance it would be beneficial to put an additional read-out protection in place, if possible.

In fact, the **shielded connection** is an (optional) mechanism to protect the confidentiality and authenticity of I^{2}C data between the OPTIGA Trust M and the host MCU. Mentioned pre-shared „Platform Binding" is used on both sides to derive session keys, mechanism is inspired by the TLS handshake (see Sect. 3.6.1) based on pre-shared secrets. Details of the OPTIGA Trust M key agreement protocol is explained in [5].

Establishing the Shielded Connection can be devided into 4 steps.

1. Generate random number on OPTIGA Trust M (slave)
2. Derive a session key (on both sides)
3. MCU (master) authentication
4. OPTIGA Trust M (slave) authentication.

Then, the encryption and authentication of commands/responses between the MCU and the SE are enabled, and the user can choose either to encrypt all communication or commands only or responses only.

5.2.3.2   SE Toolkit Functions

Based on the secure I^2C data link, the **MCU-SE pair work as a virtual security MCU** using the SE as a crypto coprocessor and a **Root-of-Trust for the IoT device**. Mentioned SE host library offers functions which work as a crypto toolbox for the IoT application. Accelerated by dedicated integrated hardware, the SE allows to achieve fast execution of functions like encryption/decryption, hash calculation and key generation with symmetric (AES-128) as well as public key algorithms (RSA-2048, ECC NIST P 256) and SHA256 hash scheme. Table 5.1provides an overview of available commands and required execution times (see [6]) incl. associated I^2C data transfer from/to host. Execution times are referring to plaintext I^2C data transmission, for a shielded connection another 5-10 ms per command are required.

Other host library functions are available, see [61]. For example, for IoT applications with a built-in usage limit for a certain service an integrated **monotonic counter** can be used. More complex use cases like using an **external secure memory** requires (see

Table 5.1   OPTIGA Trust M commands and performance

Category	Command	Function	Execution time, typ. (ms)
Secure storage	GetDataObject	Read 128 bytes	30
	SetDataObject	Write 128 bytes	55
Digital signature	CalcSign	Calculate ECC 256 signature	65
		Calculate RSA 2048 signature	310
	VerifySign	Verify ECC 256 signature	85
		Verify RSA 2048 signature	40
Data encryption	EncryptAsym	Encrypt a 127-byte message (RSA)	40
	DecryptAsym	Decrypt a 127-byte message (RSA)	315
	EncryptSym	Encrypt a 256-byte message (AES-128)	28
	DecryptSym	Decrypt a 256-byte message (AES-128)	35
Key pair generation	GenKeyPair	RSA 2048	2900
		ECC 256	55
DH key agreement	CalcSSec	ECDH	60

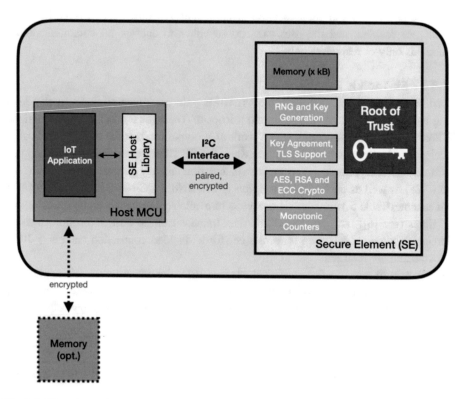

Fig. 5.4 Virtual security MCU

Fig. 5.4) can be built by a sequence of SE commands starting with the generation of symmetric key and keeping it inside the SE. Data traffic to/from external memory will be handled by the host MCU—assisted by the SE for data encryption/decryption. More complex crypto scenarios like for the **TLS protocol** are also supported, e.g., for the open-source mBedTLS implementation, see [7]. An Application Note is explaining how to implement a TLS client on the host MCU showing how to perform TLS handshake incl. mutual authentication with an echo TLS server [8].

5.2.3.3 Secure Boot and Device Integrity Checking

Main parts of the device RoT resides inside the SE. It contains provisioned data incl. the device identity (key pair and certificate) which is acting as a PKI trust anchor for external authentication and secure communication of the device within the IoT network. There is only one exception: after power-up or reset the master MCU of the device will take the lead, i.e., the host MCU. In order to prepare for reliable IoT device operation, the MCU boot process will have to make sure that it will be controlled by trusted software only. For this purpose, the **MCU boot loader** first has to verify the authenticity and integrity

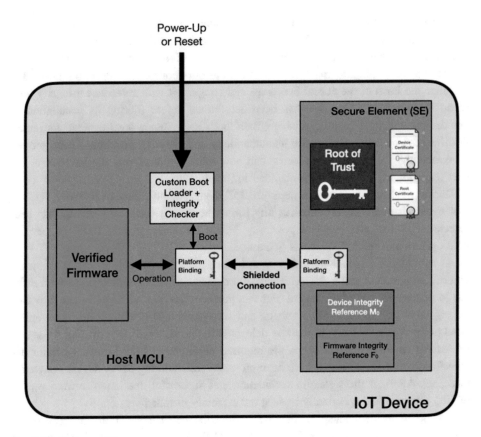

Fig. 5.5 SE Root-of-Trust

of the device software (firmware) before it will be allowed to take over and complete the startup process. Otherwise, the IoT device will reject to boot resp. terminate operation.

In a PKI-based IoT application, a digital signature will be used for this purpose (see Sect. 3.2), i.e., every authorized firmware version will be signed by the IoT business owner resp. a responsible party. This signature has to be checked by the target IoT device and will be activated in case of a positive result only. This applies to a **new firmware version** resp. to firmware updates for deployed devices in the field. But **at boot time**, there is actually no need to verify a new firmware version. Instead, we just have to double-check if the actual firmware is **identical with the known and verified firmware** version which is already installed on the IoT device. This can be done with a secure hash function which has to matching values of both firmware versions.

Anyway, in case of this SE-centered approach, the host MCU can **delegate firmware verification work to the SE**. As a consequence, the MCU boot loader has to cooperate with the SE, so the default **MCU boot loader must be replaced** by a SE-centric **custom**

boot loader (see Fig. 5.5). The custom boot loader, after device power-up or reset, will first have to initialize the shielded connection to the SE, see Sect. 5.2.3.1. Now the SE is available as a secure co-processor and to assist verification of the actual MCU firmware. For secure boot, the "integrity checker" of the custom boot loader will let the host MCU calculate the **hash of the actual firmware** and compare it with **reference integrity hash value F_0** (see Fig. 5.5) which has been determined before (during its installation on the device) and kept securely inside SE. If both hash values are identical, the actual firmware can be activated because its authenticity and integrity has already been proven. Consequently, the custom boot loader can now release control, and the IoT device can start operation managed by the embedded IoT application.

During field operation of the deployed IoT device, same firmware integrity verification mechanism can be **repeated at any time** whenever a tampering attempt has been detected—not just after reset or power-up. Based on this approach, 100% tamper-free field operation of the IoT device is guaranteed.

As mentioned, a **secure firmware update** function for deployed devices is a key IoT security requirement and a powerful tool. On device side, the installed firmware version has to manage its own replacement. For this purpose, the host MCU will have to extract the signature of a new firmware version and verify that it has been signed by an authorized party. Again, this crypto work is delegated to the SE which is providing the secure processing environment and keeps the required public key of the firmware creator and associated certificate(s) which have been pre-provisioned during SE or device production (see Sect. 4.3.9). If the signature verification was successful, the new firmware version will be activated and will finally replace the currently installed version.

On top of its own integrity, the IoT application can benefit from the SE-centered approach to ensure **integrity of the complete IoT device**, recall Sect. 4.2.3. For this purpose, besides embedded software also system components and configuration registers are being determined by a process call "integrity measurement". The initial reference value M_0 is reflecting the deployment status of the device (see Sect. 4.3.9). Same measurement procedure can be initiated by the IoT application or remotely by the IoT operator whenever required or in case of doubt.

References

1. STMicroelectronics. (2022). Ultra-low-power Arm Cortex-M33 32-bit MCU+TrustZone. Retrieved June 10, 2022, from https://www.st.com/en/microcontrollers-microprocessors/stm32u585ai.html.
2. Vishay Semiconductors. (2022). VEML6030 High Accuracy Ambient Light Sensor With I2C Interface. Retrieved June 19, 2022, from https://www.vishay.com/optical-sensors/list/product-84366/.

3. Infineon. (2022). OPTIGA™ TRUST M SLS32AIA. Retrieved June 19, 2022, from https://www.infineon.com/cms/en/product/security-smart-card-solutions/optiga-embedded-security-solutions/optiga-trust/optiga-trust-m-sls32aia.
4. GitHub. (2022). Infineon Optiga Trust M. Retrieved June 19, 2022, from https://github.com/Infineon/optiga-trust-m.
5. Infineon. (2022). IFX I2C Protocol Specification Retrieved June 19, 2022, from https://github.com/Infineon/optiga-trust-m/blob/develop/documents/Infineon_I2C_Protocol_v2.03.pdf.
6. Infineon. (2022). OPTIGA TRUST M SLS32AIA Datasheet. Retrieved June 24, 2022, from https://github.com/Infineon/optiga-trust-m/blob/develop/documents/OPTIGA_Trust_M_Datasheet_v3.10.pdf.
7. GitHub. (2022). Mbed TLS. Retrieved June 24, 2022, from https://github.com/Mbed-TLS.
8. GitHub. (2022). TLS client implementation using MbedTLS crypto library with OPTIGA™ Trust M. Retrieved June 24, 2022, from https://github.com/Infineon/mbedtls-optiga-trust-m.

Glossary

Description of common terms and acronyms related to technical IoT, cellular network and security topics.

Note: **bold** items indicate terms with its own description in this glossary.

Term	Acronym	Description
3rd generation partnership project	3GPP	Global standardization body for cellular mobile telecommunication protocols like **LTE**
Advanced encryption standard	AES	Algorithm used for **symmetric cryptography**
Application-specific standard products	ASSP	Standard chip component addressing a specific electronic application (e.g., a cellular network interface). In contrary to customer-specific ASICs, an ASSP is designed to be used by different customers
Asymmetric cryptography		See **Public-key cryptography**
AT command interface		Standard network programming interface (API) used for interaction with a **modem**
Application programming interface	API	Specific software connection offering services to other software modules
Authentication		Process of verifying the identity of a person of networked object
Bill of material	BOM	List of the parts, and the quantities of each needed to manufacture an end product

© The Editor(s) (if applicable) and The Author(s), under exclusive license to Springer Nature Switzerland AG 2022
K. Heins, *Trusted Cellular IoT Devices*, Synthesis Lectures on Engineering, Science, and Technology, https://doi.org/10.1007/978-3-031-19663-8

Term	Acronym	Description
Block cipher		Asymmetric cryptographic algorithm that encrypts messages by breaking them down to fixed-size blocks
Brute force attack		A method that tries to decipher an encrypt a message with *guessed* keys
Coverage enhancement	CE	Methods to improve **LTE-M** and **NB-IoT** cellular network reach using repeated data transmissions and **HARQ** error correction
Cellular IoT	CIoT	IoT based on cellular network technology
Central processing Unit	CPU	Executes instructions of a computer program. Core of a **microcontroller (MCU)**
Certificate		Short for **public-key certificate**
Certification authority	CA	Trusted 3rd-party entity that issues **digital certificates** for a user group
Chain of trust		A layered structure of certificates/signatures (based on a **trust anchor**) assuring the trustworthiness of other elements within the structure. Validity of each layer is guaranteed by the previous layer to create a chain
Checksum		A small, easy to calculate digital value representing the larger original file, e.g., to be used to verify its integrity
Cipher		Alternate word for a cryptographic algorithm
Ciphertext		Encrypted original data (**plaintext**)
Common criteria	CC	Standard for security evaluations
Common criteria evaluation assurance level	CC EAL	Certified security confidence level following a completed product evaluation according to **Common Criteria** rules
Countermeasures		Process or implementations that can prevent or mitigate the actions of a threat or an attack
Cyberattack		Offensive maneuver against computer systems, networks, infrastructures

Term	Acronym	Description
Decryption		Process of using a cryptographic algorithm to convert encrypted data (**ciphertext**) back into original data (**plaintext**)
Data encryption standard	DES	Algorithm used for **symmetric cryptography**. Not used any more
Diffie-hellmann	DH	Handshake protocol and algorithm to create a shared secret key (session key) for a secure communication channel—based by exchanging public information between both parties
Denial-of-service attack	DoS	**Cyberattack** to make a network resource unavailable, e.g., by overloading it with superfluous requests
Differential power analysis		A type of **Side-Channel Attack (SCA)** based on analyzing power consumption variations of an electronic circuit performing crypto operations involving confidential keys
Digital certificate		See **Public-key certificate**
Digital signature		Mathematical scheme for verifying the authenticity and the integrity of digital messages or documents
Digital signature algorithm	DSA	Algorithm used for **public-key cryptography**
Edge computing		Distributed computing paradigm that brings computation and data storage closer to the sources of data. IoT follows this idea
eDRX	extended discontinuous reception	**LTE** power saving feature
Elliptic curve cryptography	ECC	Algorithm used for **public-key cryptography** based on elliptical curve constraints
Elliptical curve diffie-hellman	ECDH	Combination of Elliptical curve cryptography and **Diffie-Hellman** key exchanges to generate a shared secret

Term	Acronym	Description
Elliptical curve diffie-hellman ephemeral	ECDHE	**ECDH** done with temporary (ephemeral) keys. After the secret is used, it is destroyed, along with the temporary key pairs. This type of temporal secret is fundamental to achieving **Perfect Forward Secrecy**
Encryption		Process of using a cryptographic algorithm to convert original data (**plaintext**) into incomprehensible data (**ciphertext**)
End-to-end encryption	E2EE	Communication system where only the communicating users can read the messages—using cryptographic keys needed to decrypt the conversation which are accessible by communication users only
evolved NodeB	eNodeB (eNB)	LTE base station
Entropy		Quality of a random number (unpredictability)
Fault injection attack	FI	Attack method to change behavior of a system by exposing it with extreme conditions, e.g., clock glitches or higher temperature or supply voltage
Federal information processing standards	FIPS	Standards set by the US government for data protection
Firmware		Embedded software stored in non-volatile memory of a computing device
Firmware over-the-air	FOTA	Technical approach for remote updating the **firmware** of an (IoT) device via wireless network
General data protection regulation	GDPR	EU law on data protection and privacy in European Union
General packet radio service	GPRS	2G/3G cellular mobile data standard
GitHub		GitHub.com is a hosting platform for software development, commonly used for open-source software projects where users can download latest versions and documentation
Global positioning system, global navigation satellite system	GPS, GNSS	Satellite-based radio navigation systems

Term	Acronym	Description
Hands-on attack		Security attack where attackers have physical access to a crypto system
Hardware security module	HSM	Tamper-resistant computing device for management of cryptographic keys
HARQ	Hybrid automatic repeat request	LTE process combining data retransmission and error correction
Hash function		Mathematical function used to map data of arbitrary size to fixed-size values, e.g., a message digest
Hash function, secure		A **hash function** respectively hash algorithm is called secure if two conditions are met: (1) it works only one way, i.e. it is computationally infeasible to find a message that corresponds to a given message digest (2) it is impossible to find two different messages that produce the same message digest
Industrial IoT	IIoT	IoT solutions aiming at industrial applications
Integrated circuit	IC	Active electronic component integrating multiple functional elements into one single semiconductor-based device
Integrity		Assurance that (transmitted) data have not been altered, modified, or replaced
Integrity measurement		Process to determine functional system details in order to verify its integrity, used for **secure boot** function
International mobile subscriber identity	IMSI	Unique identifier for mobile user, assigned by MNO in SIM respectively eSIM
Intrusion detection system	IDS	Dedicated device or software application that monitors an IT system for unauthorized access or malicious activity
IoT cloud		On-demand IT services to facilitate IoT devices including collection, analysis, processing of IoT data

Term	Acronym	Description
IoT ecosystem		Ingredients of an IoT system consisting of sensors/actuators, an embedded application, network connectivity, server-based application software and security measures
Key		A parameter, such as a private key, public key, secret key or session key that is used in cryptographic functions
Key pair		Corresponding public and private keys, used for **Public-key cryptography**
Key management		Management of cryptographic keys in a cryptosystem including the generation, exchange, storage, use, crypto-shredding (destruction) and replacement of keys
Key space		Collection of all possible keys in a cryptosystem
Key schedule		An algorithm that creates subkeys in cipher blocks within a given **key space**
Leaf certificate		Is a **public-key certificate**, but in this case, it has not been issued by a **CA** but by a trusted end device and confirms that the certified has been created by this device
Logical attack		Type of **cyberattack** that is performed remotely, i.e., by software via an vulnerable **API**
(Long Range)	LoRa	Proprietary, non-cellular **LPWAN** technology, in unlicensed spectrum
Low-power wide area network	LPWAN	Category of wireless networks technologies incl. **NB-IoT** and **LTE-M**
Long-term evolution	LTE	Standard for wireless broadband communication, developed by **3GPP**
Long-term evolution—machine (Type communication)	LTE-M(TC)	Cellular LPWAN technology based on **3GPP** LTE standard
Machine-to-machine	M2M	Direct communication between network devices

Term	Acronym	Description
Man-in-the-middle attack		**Cyberattack** where the attacker sits between two communication partners and eavesdrops or alters transmitted data
Maximum coupling loss	MCL	Maximum signal loss that a wireless system can tolerate and still be operational
Message authentication code	MAC	Message tag created from plaintext using a **MAC algorithm** and symmetric key that ensures authentication and data integrity
MAC algorithm		Common algorithms are HMAC-MD5, HMAC-SHA-1 and HMAC-SHA-512
Message digest		Output value of a **hash function**
Microcontroller	MCU	Single-chip computer containing one or more processor cores along with on-chip memory and programmable input/output peripherals
MNO, MVNO	Mobile network operator, mobile virtual network operator	Wireless communications services provider
Modem (=modulator-demodulator)		Hardware subsystem to convert digital data into a specific format for analog transmission
Message queueing telemetry transport	MQTT	TCP/IP Application Layer lightweight, publish-subscribe network protocol
National institute of standards and technology	NIST	Physical science lab and US federal agency responsible for technical administration and standardization
Narrowband IoT	NB-IoT	Cellular LPWAN technology based on **3GPP** LTE standard
Near-field communication	NFC	A protocol for short-range and low-speed communication of two electronic devices over a distance of 4 cm or less. NFC devices can be used for identity or payment tokens
Nonce		An arbitrary number or bit string used only once

Term	Acronym	Description
Penetration testing	Pentesting	Simulated **cyberattack** against an **embedded system**, performed to evaluate the security and to identify weaknesses (vulnerabilities)
One-time programmable memory	OTP	Read-only, not-updatable memory which can be written only once
Perfect forward secrecy		Protects past sessions against future compromises of secret keys or passwords
Physical unclonable function	PUF	A physical object that serves as a unique identifier, typically for semiconductor products
Plaintext		Original unencrypted data, also called cleartext, i.e., before is becomes **cyphertext** via **encryption**
Platform security architecture	PSA	IoT security scheme originated from company Arm Holding
Private key		In symmetric cryptography, the private key is equivalent with the secret key (shared key). In asymmetric cryptography, the private key is the secret half of the public/private key pair
Printed circuit board	PCB	Sandwich-structured board for assembly and interconnection of electronic components
Pseudo-random number	PRN	Numbers that seem random but are actually determined by specific function and seed value. PRNs are created by a PRNG (PRN Generator)
Public land mobile network	PLMN	Unique ID for country and network operator
Public-key certificate		Electronic document used to prove the ownership of a public key. For this purpose, typically, a **digital signature** of a **Certification Authority (CA)** is used as a **trust anchor**
Public key infrastructure	PKI	A complete set of roles, policies, and procedures needed to create manage, distribute, use, store, and revoke digital certificates for cryptographic keys and manage a public-key cryptosystem

Term	Acronym	Description
Power save mode	PSM	**LTE** feature allowing **modems** and IoT devices to enter idle mode and lowest power consumption level
Public-key cryptography	PK crypto	Asymmetric cryptographic system that uses a pair of keys (a public and a private key), one for the encryption of plaintext and the other for decryption of ciphertext
Radio access technology	RAT	Connection method for a radio-based communication network, e.g., LTE, Bluetooth, WiFi
Random access	RA	Cellular network procedure initiated by user device to apply for data transfer
Radio ressource control	RRC	Cellular network protocol used between IoT device and Base Station
Real-time clock	RTC	Electronic circuit (integrated in a **MCU**) providing exact data and time
Rivest–shamir–adleman	RSA	Algorithm used for **public-key cryptography**
Reference signal received power	RSRP	Measured value of received power of the LTE reference signals
Reference signal received quality	RSRQ	Measured value of received quality of the LTE reference signals
Registration authority	RA	Trusted authority of a **PKI** which is verifying a **certificate** request and defines its content
Root CA		Top-level certification authority issuing certificates for next CA levels in a hierarchical CA infrastructure. Works as a **trust anchor**
Root of trust	RoT	Hardware-based secure foundation of a cryptosystem offering ultimate tamper resistance for a **Chain of Trust** used for device authentication, secure data transmission, secure boot, etc.
Secret key		A shared key used for encryption and decryption in **symmetric cryptography**

Term	Acronym	Description
Secure element		A **tamper-resistant** component used to securely store sensitive data, keys, and to execute cryptographic functions and secure services
Secure hash algorithm	SHA	A **hash** algorithm that creates a *unique* value for each input
Session key		A key used only valid during a single communication session between two parties
Short message service	SMS	Text messaging service of cellular networks
Side-channel attack	SCA	A security attack that takes advantage of physical information leakage from a cryptosystem (e.g., a semiconductor circuit) in order to extract confidential keys or secret information. Common attacks are exploiting timing information, power consumption and electromagnetic emission during cryptographic operations
SigFox		Proprietary, non-cellular **LPWAN** technology, in unlicensed spectrum
Signal to interference plus noise ratio	SINR	Quality indicator for a wireless connection
Smart grid		Technical infrastructure to facilitate efficient energy supply, distribution and consumption services
Smart meter		IoT device for measurement of energy consumption. Part of the **smart grid**
Subscriber identification module	SIM	Smart card provided by network provider to authenticate user device
Symmetric cryptography		Symmetric cryptographic system that uses the same (secret) key for both the encryption of plaintext and the decryption of ciphertext
Tamper-evidence		In contrast to **tamper-resistance**, tamper-evident devices only detect and indicate tampering attempts (but do not prevent them)
Tamper-resistance		Applied hard- and software methods preventing unauthorized use or access to internal data of a device ("tampering")

Term	Acronym	Description
Transport layer security	TLS	Standard secure communication protocol creating an encrypted link between a web server and a browser. Its predecessor is known as Secure Sockets Layer (SSL)
True random number	TRN	A hardware device that generates random numbers from a *physical* process rather than from an algorithm. TRNs are created by a TRNG (TRN Generator). See also **Pseudo-Random Number (PRN)**
Trusted execution environment	TEE	A TEE is an environment for executing authorized code only—and ignores all other attempts
Trusted third party	TTP	A TTP provides services or facilitates interactions for a community. A typical TTP is a **certification authority (CA)** in a **PKI infrastructure (PKI)**
Trust anchor		A trust anchor is an authoritative entity that is trusted per se by a user group in a **public-key infrastructure**, e.g., a signed digital **certificate** by a **root CA** ("root certificate") in a **Chain of Trust**
TrustZone		Security **CPU** architecture (by ARM) for isolating trusted software in its own virtual processing environment
Vulnerabilities		Weaknesses of an **embedded system**, potentially to be used by **cyberattacks**
Watchdog timer		A watchdog timer is a simple countdown timer that is used to reset a **MCU** after a specific interval of time—in order to increase system reliability
WiFi		A family of wireless network protocols based on the IEEE 802.11 family of standards. Commonly used for local area networking of devices and Internet access

Printed in the United States
by Baker & Taylor Publisher Services